Green Organic Chemistry

Strategies, Tools, and Laboratory Experiments

Kenneth M. Doxsee
University of Oregon

James E. Hutchison
University of Oregon

Annie —
Looking forward
to working with you
on green chemistry !
in the future
Best —

THOMSON

BROOKS/COLE

Australia • Canada • Mexico • Singapore • Spain • United Kingdom • United States

Printed in the United States of America
1 2 3 4 5 6 7 07 06 05 04 03

Printer: Globus Printing Company, Inc.

ISBN: 0-534-38851-5

For more information about our products, contact us at:
Thomson Learning Academic Resource Center
1-800-423-0563

For permission to use material from this text, contact us by:
Phone: 1-800-730-2214
Fax: 1-800-731-2215
Web: http://www.thomsonrights.com

Asia
Thomson Learning
5 Shenton Way #01-01
UIC Building
Singapore 068808

Australia/New Zealand
Thomson Learning
102 Dodds Street
Southbank, Victoria 3006
Australia

Canada
Nelson
1120 Birchmount Road
Toronto, Ontario M1K 5G4
Canada

Europe/Middle East/South Africa
Thomson Learning
High Holborn House
50/51 Bedford Row
London WC1R 4LR
United Kingdom

Latin America
Thomson Learning
Seneca, 53
Colonia Polanco
11560 Mexico D.F.
Mexico

Spain/Portugal
Paraninfo
Calle/Magallanes, 25
28015 Madrid, Spain

Preface to the First Edition

As we have developed the green organic chemistry program at the University of Oregon, it has become apparent to us that there is an urgent need and rapidly growing demand for green chemistry laboratory materials. As the field of green chemistry continues to undergo rapid growth and change, so too will this text. Future editions will incorporate new green chemical strategies and discussion materials, address new questions, and present new experiments designed to bring to life the evolving concepts and practice of modern organic chemistry.

The first section of this text is devoted to a description of green chemistry and the tools and strategies used in its implementation. This material is an essential complement to the experimental descriptions that follow. The experiments allow investigation of green organic chemical concepts in the laboratory setting. That said, the experiments are not essential to an analysis of the first, descriptive half of the text, and it is our hope that this text will prove useful to those interested in learning about the principles of green chemistry in general.

We have made the decision not to present the general "techniques" and "procedures" sections – melting point determination, recrystallization, spectroscopy, etc. – typically found in organic chemistry laboratory texts. We anticipate that this text will be used in conjunction with a text or manual containing descriptions of these procedures and techniques.[†] We have not written the experimental procedures with reference to any one particular such text, however, as we believe this will allow the greatest individual latitude in selection of a companion text suitable to your local setting. As we continue to work toward later editions, comments from you, the reader and user of this edition, about this decision will be valuable to us. Would you prefer this text to be a "stand-alone" text, or are you content to "package" it with a companion "techniques" text?

[†] Examples of such texts include R. J. Fessenden, J. S. Fessenden, & P. Feist, *Organic Laboratory Techniques*, 3rd Ed.; Brooks/Cole: Pacific Grove, CA, 2001; J. R. Mohrig, C. N. Hammond, P. F. Schatz, & T. C. Morrill, *Techniques in Organic Chemistry*; W. H. Freeman & Co.: New York, 2003; J. W. Zubrick, *The Organic Chemistry Lab Survival Manual: A Student Guide to Techniques*, 5th Ed.; John Wiley & Sons: New York, 2000.

Nineteen experiments are included in this first edition. These experiments were designed to complement the content of the typical undergraduate organic chemistry sequence while illustrating the principles and strategies of green organic chemistry. We have presented the experiments in an order we have found useful, but they may be used in any desired sequence. Later editions of the text will include a wider range of experiments, including a number that are under active development. Given the nature of this first edition, here too is an opportunity for you to influence its further development. Please let us know of your personal experiences with the experiments, and feel free to suggest changes, as well as topics for experiments you would like to have available. In anticipation of one suggestion, we note here our intention to include in later editions microscale procedures in addition to the larger scale procedures reported in this edition.

KMD

JEH

February 11, 2003

All experiments contained in this laboratory manual have been performed safely by students in college laboratories under the supervision of the authors. However, unanticipated and potentially dangerous situations are possible due to failure to follow proper procedures, incorrect measurement of chemicals, inappropriate use of laboratory equipment, or other reasons. The authors and the publisher hereby disclaim any liability for personal injury or property damage claimed to have resulted from use of this laboratory manual.

Acknowledgments

The design and testing of the experiments presented in this text relied on the qualified and enthusiastic efforts of a group of talented co-workers to whom the authors are profoundly grateful: Dr. Scott Reed, Lauren Huffman, Marvin Warner, Dr. Robert Gilbertson, Gary Succaw, Lallie McKenzie, Kathryn Parent and Gerd Woehrle all made important contributions toward the experiments described in this text. We are deeply appreciative of the efforts of Dr. Leif Brown, who was the first to "run this experiment on a large scale," bringing this curriculum to our entire organic laboratory course and providing invaluable experimental troubleshooting in the process. We also thank Leif for many contributions to the experiment descriptions presented in this text. Dr. John Thompson at Lane Community College made a number of excellent suggestions for improving the labs as a result of his in-class testing of most of the experiments at LCC.

This project would not have been possible without the support of the chemistry department and the university administration. In particular, we thank Prof. David Tyler and Prof. David Johnson whose support in launching the program was invaluable. A number of our colleagues, including the many participants of our annual Green Chemistry in Education Workshops, have made important contributions. In particular, we thank Dr. Paul Anastas, Dr. Mary Kirchhoff, Dr. Julie Haack, Dr. Lauren Heine, Dr. Robert Hembre and Dr. Richard Wolf for their numerous contributions and enduring support of our efforts. We thank Dr. Haack for her continuing contributions to the development of green chemistry educational materials within the department and her support of this program as our Assistant Department Head. We also acknowledge the administrative support provided by Kristi Mikkelsen in the preparation of these materials and the organization of our annual workshops. Mary Dricken, Gary Nolan, and Sandi Smith (members of our chemistry department staff) all contributed to the remarkably smooth running of the laboratory program throughout its development, as well as the annual Green Chemistry in Education Workshop held at the University of Oregon each summer.

We are pleased to acknowledge gratefully the support of our efforts by the National Science Foundation, the Alice C. Tyler Perpetual Trust, the Green Chemistry Institute of the American Chemical Society, the Environmental Protection Agency and the American Chemical Society.

Finally, we express our appreciation to all the students (and their teaching assistants) of Chemistry 337 and Chemistry 338 at the University of Oregon, particularly those who either volunteered to be or put up with becoming guinea pigs in the testing and optimization of these new experiments, and last, but by no means least, the technical wizardry of Ryan Stasel, who late in the process recovered the electronic version of this text from a malfunctioning computer.

KMD

JEH

February 11, 2003

Preface

This text is the product of six years of work toward the development of a green organic chemistry laboratory program at the University of Oregon, culminating in the complete replacement of our organic chemistry laboratory sequence with a greener curriculum. When students learn organic laboratory chemistry at the University of Oregon, they learn green organic chemistry. While becoming versed in the essentials of organic chemical theory and practice representing an essential feature of the organic chemistry laboratory experience, they at the same time acquire the tools to assess the health and environmental impacts of chemical processes and the strategies to improve them. The conception, design, and implementation of our green organic chemistry laboratory program followed a rather complex path, and we feel that our experiences and thought processes, outlined in the following paragraphs, may be helpful to others contemplating the conversion to a greener curriculum.

Green Organic Chemistry in the Laboratory: Practical Solution & Golden Opportunity

Many of us have struggled with the problem of how to modify our organic laboratory curriculum to provide a modern organic chemical experience, reduce the amount of waste generated, and make the lab safer for our students. Now-traditional approaches to minimizing waste and improving safety in the organic laboratory involve use of microscale experimentation and provision of increased ventilation. We initially pursued this approach, renovating a laboratory, at considerable expense, to allow each student to perform his/her microscale work within a fume hood. Although these changes reduced the amount of waste and provided a safer working environment, we were frustrated by several major disadvantages. A significant practical problem was that the laboratory's capacity of eighteen students required us to schedule night and weekend laboratory sessions for some students. The fume hoods were expensive to install and to operate, and created a noisy workspace with restricted sight lines. More fundamentally, we felt that our exclusive reliance on microscale methods was not adequately preparing our students for work at the larger scale found in our research laboratories or in industry.

In the process of addressing these issues we asked the question, *"Can we use 'green' chemistry methods to teach organic laboratory skills and chemical concepts, using standard taper glassware, on the bench top?"* Our goal was to use the latest advances in green chemistry research to develop a new laboratory curriculum that would decrease our reliance on fume hoods, produce a less hazardous waste stream, and improve safety. The fact that green methods eliminate hazards rather than simply attempting to prevent exposure through (fallible) safety procedures made it ideal for implementing in the teaching laboratory setting. As we began to develop the idea, other benefits quickly became evident. Green chemical methods achieve hazard reduction at all reaction scales, permitting the introduction of larger-scale experimentation, using standard laboratory glassware. Green chemistry offers the opportunity to upgrade comprehensively the organic lab curriculum, replacing many classic but perhaps timeworn experiments. Green chemistry provides a unique context for more detailed discussions of chemical hazards and the effects of chemicals on human health and the environment.

We rapidly discovered there were few existing green experiments suitable for the organic chemistry teaching laboratory. This realization led us to search for existing experiments that could be modified to be greener and for new reports of green methods that could be adapted for use in the teaching laboratory. We established the following criteria for judging the suitability of potential new experiments.

- Reduces laboratory waste and hazards.
- Illustrates green chemical concepts.
- Teaches modern reaction chemistry and techniques, in parallel with lecture course.
- Provides a platform for discussion of environmental issues in the classroom.
- Can be accomplished by students during a typical organic laboratory period.
- Uses inexpensive, greener solvents and reagents.
- Is adaptable to both macroscale and microscale methods.

In cases where experiments were to be modified, we adopted a set of principles to guide the process of modifying the experiments.

- Eliminate (minimize) hazardous solvents whenever possible, both as reaction media and in solvent-dependent separations.
- Identify and use the most benign reagents possible, using mild, more selective reagents to replace traditional, overly reactive reagents.

- Develop and use efficient reaction chemistry – including catalysis, atom economy, etc.

Over the past five years, we have developed and student-tested a number of experiments based on these criteria, and we have found that green organic laboratory is a practical mechanism for teaching core skills and concepts in a safer and environmentally-friendlier manner. In the process of testing the curriculum, we have discovered a number of unexpected benefits. Since students are regularly required to evaluate the chemical hazards associated with a method and learn to take appropriate precautions, they learn to assess hazards critically rather than simply assuming that all chemicals are "toxic." This alters their (often) negative perception of chemistry.

Another significant benefit derives from students being exposed to the strategies and methods we used to make the experiments green. While one can simply use green chemical experiments as the mainstay of the organic laboratory program without ever discussing green chemistry, to do so is to miss a rare opportunity to change the way that students view chemistry and their perceived ability to contribute to it. For each experiment, we show the students how the new green method differs from traditional methods and describe the process we or others used in modifying the experiment. Through repetitive exposure to the process of analyzing reaction conditions and finding alternative methods, students learn to assess the conditions for hazards and likelihood of exposure/release, identify new methods that allow reduction of the hazards, explore how alternative methods can affect the course and rate of reactions, and assess the broader impact of the overall process on safety and the environment. This aspect of the curriculum is especially empowering to students. They realize that, armed with this knowledge, they can help disseminate better and more sustainable chemical practices throughout academic and industrial settings, acting as *ambassadors of green chemistry.*

Although our green organic lab curriculum is still evolving, we are excited about its practical advantages and the unique opportunities it offers for promoting chemistry as a responsible science. By introducing state-of-the-art green chemical methods, objectively discussing chemical hazards and preparing students to help shape the future of chemical practices, we can demonstrate to students that chemistry can be used to solve, rather than cause, environmental problems. In so doing, we positively influence students' opinions about chemistry and encourage them to take an active part in shaping the future of chemistry in our society.

Brief Table of Contents

Contents

Laboratory Experiments

Graphical Abstracts for the Experiments

---

1. SOLVENTLESS REACTIONS: THE ALDOL CONDENSATION (P. 115 - 119)

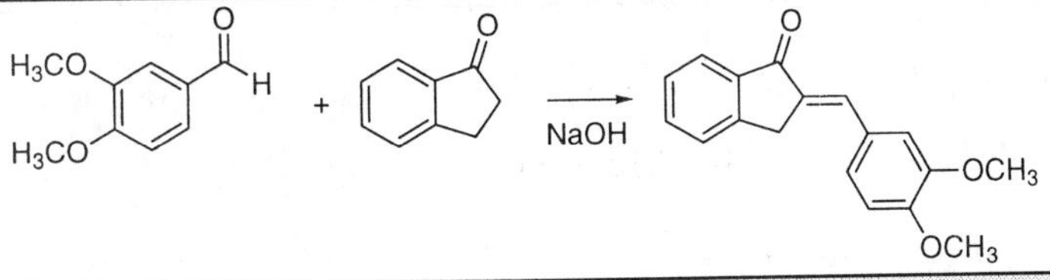

Concepts/techniques:
Carbonyl chemistry; the aldol reaction; melting points of solids and mixtures; recrystallization.

Green chemistry messages:
Solventless reactions between solids, atom economy.

---

2. BROMINATION OF AN ALKENE: PREPARATION OF STILBENE DIBROMIDE (P. 120 - 124)

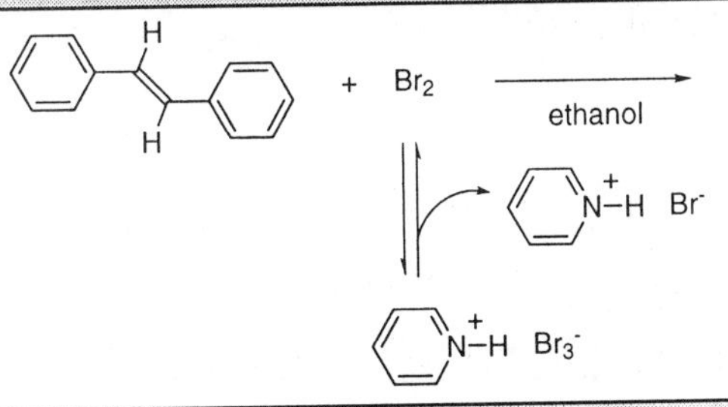

Concepts/techniques:
Halogenation of alkenes; reactions at elevated temperature; vacuum filtration; melting point determination.

Green chemistry messages:
Safer solvents; safer reagents.

---

3. A GREENER BROMINATION OF STILBENE (P. 125 - 128)

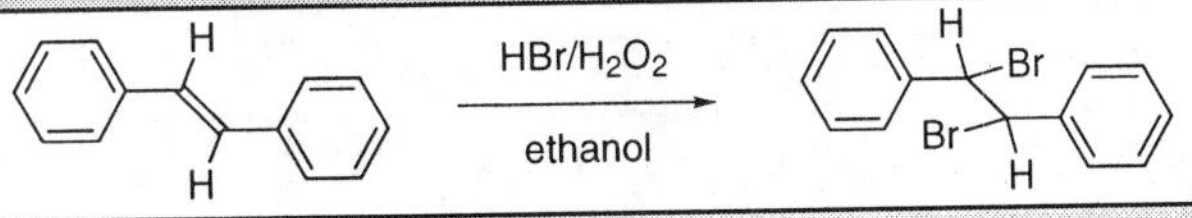

Concepts/techniques:
Halogenation of alkenes; reactions at elevated temperature; vacuum filtration; melting point determination.

Green chemistry messages:
Safer solvents; safer reagents; atom economy; relative nature of "green."

4. PREPARATION AND DISTILLATION OF CYCLOHEXENE (P. 129 - 134)

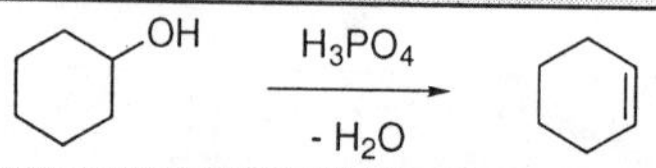

Concepts/techniques:

Dehydration of alcohols; multi-step syntheses; liquid-liquid extraction; drying agents; simple and fractional distillation; boiling point determination; infrared (IR) spectroscopy.

Green chemistry messages:

Safer reagents; solvent-free synthesis.

5. SYNTHESIS AND RECRYSTALLIZATION OF ADIPIC ACID (P. 135 - 141)

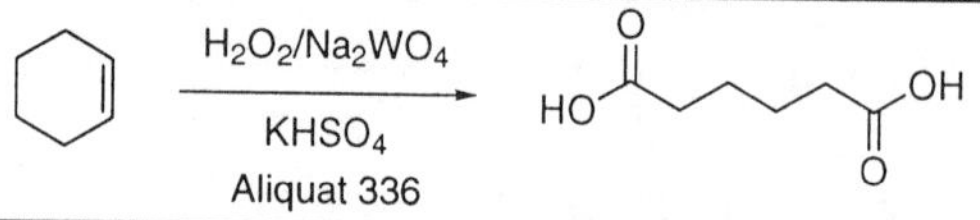

Concepts/techniques:

Oxidative cleavage of an alkene C=C bond; phase transfer catalysis; recrystallization; melting-point determination; polymer chemistry.

Green chemistry messages:

Catalysis; alternative reaction media; reuse of reagents.

6. OXIDATIVE COUPLING OF ALKYNES: THE GLASER-EGLINTON-HAY COUPLING (P. 142 - 151)

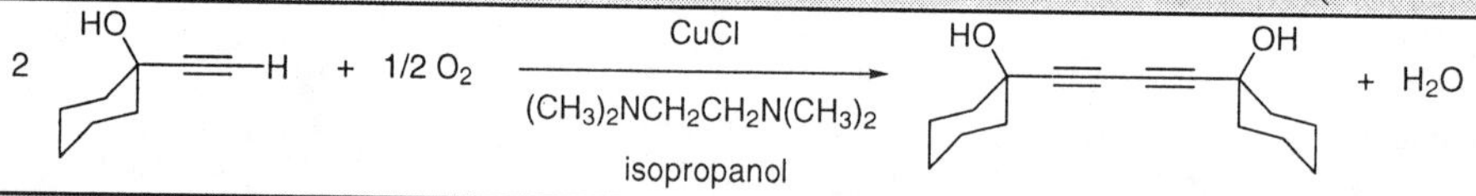

Concepts/techniques:

Oxidative coupling of alkynes; decolorization; thin-layer chromatography (TLC); infrared (IR) spectroscopy; preparation of KBr pellets.

Green chemistry messages:

Catalysis; alternative solvents; mild reagents (molecular oxygen).

7. GAS PHASE SYNTHESIS, COLUMN CHROMATOGRAPHY AND VISIBLE SPECTROSCOPY OF 5,10,15,20-TETRAPHENYLPORPHYRIN (P. 152 - 158)

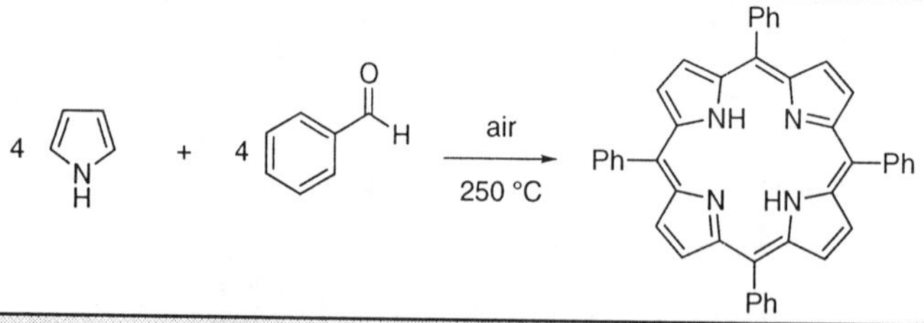

Concepts/techniques:

Synthesis of porphyrins; electrophilic aromatic substitution; gas-phase reactions; column chromatography; UV-visible spectroscopy; thin-layer chromatography.

Green chemistry messages:

Solvent-free reactions; avoidance of corrosive reagents; oxidation by air.

8. MICROWAVE SYNTHESIS OF 5,10,15,20-TETRAPHENYLPORPHYRIN (P. 159 - 162)

Concepts/techniques:

Electrophilic aromatic substitution; column chromatography; visible spectroscopy; thin-layer chromatography.

Green chemistry messages:

Solvent-free reactions; solid-supported synthesis; microwave heating of reaction mixtures; safer solvents (for chromatography).

9. METALLATION OF 5,10,15,20-TETRAPHENYLPORPHYRIN (P. 163 - 166)

Concepts/techniques:

Coordination chemistry; visible spectroscopy.

Green chemistry messages:

Safer solvents; elimination of need for heating.

10. MEASURING SOLVENT EFFECTS: KINETICS OF HYDROLYSIS OF *TERT*-BUTYL CHLORIDE (P. 167 - 173)

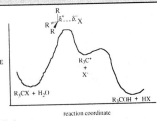

Fig. 1

Concepts/techniques:

S_N1 reactions; solvent effects; chemical kinetics; titration.

Green chemistry messages:

Alternative solvent selection – recognition of impact of solvent choice on reaction rate.

11. MOLECULAR MECHANICS MODELING (P. 174 - 181)

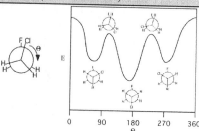

Concepts/techniques:

Computational methods; molecular mechanics calculations; molecular modeling.

Green chemistry messages:

Design of safer products; synthetic efficiency.

12. ELECTROPHILIC AROMATIC IODINATION (P. 182 - 188)

Concepts/techniques:

Electrophilic aromatic substitution; recrystallization; melting point determination.

Green chemistry messages:

Safer and easier to handle reagents and solvents, more selective reagents.

13. PALLADIUM-CATALYZED ALKYNE COUPLING/INTRAMOLECULAR ALKYNE ADDITION: NATURAL PRODUCT SYNTHESIS (P. 189 - 196)

Concepts/techniques:

Alkyne chemistry (acetylides, addition reactions); substitution reactions; organometallic chemistry; air-sensitive compounds and techniques for handling them.

Green chemistry messages:

Use of water as a solvent; catalysis; efficient synthetic routes.

14. RESIN-BASED OXIDATION CHEMISTRY (P. 197 - 200)

Concepts/techniques:

Oxidation chemistry; solid-supported reagents.

Green chemistry messages:

Nontraditional reagents and conditions; recyclable reagents.

15. CARBONYL CHEMISTRY: THIAMINE-MEDIATED BENZOIN CONDENSATION OF FURFURAL (P. 201 - 205)

Concepts/techniques:

Reactions of carbonyl compounds; oxidation chemistry; carbon skeleton rearrangements.

Green chemistry messages:

Safer and easier to handle reagents and solvents.

16. SOLID-STATE PHOTOCHEMISTRY (P. 206 - 210)

Concepts/techniques:

Photochemistry; crystal engineering; recrystallization.

Green chemistry messages:

Photochemical methods; solid-state (solvent-less) reactions; atom economy of addition reactions; selective synthesis; alternative energy sources.

17. APPLICATIONS OF ORGANIC CHEMISTRY: PATTERNING SURFACES WITH MOLECULAR FILMS (P. 211 - 224)

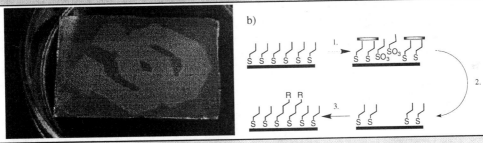

Concepts/techniques:

Organic materials chemistry; surface chemistry, self-assembled monolayers (SAMs), controlling surface properties with organic thin films; mechanical and chemical patterning of surfaces.

Green chemistry messages:

Design of processes requiring less material ("dematerialization"); low-temperature, high-efficiency processes; benign solvents.

18. THE FRIEDEL-CRAFTS REACTION: ACETYLATION OF FERROCENE (P. 225 - 230)

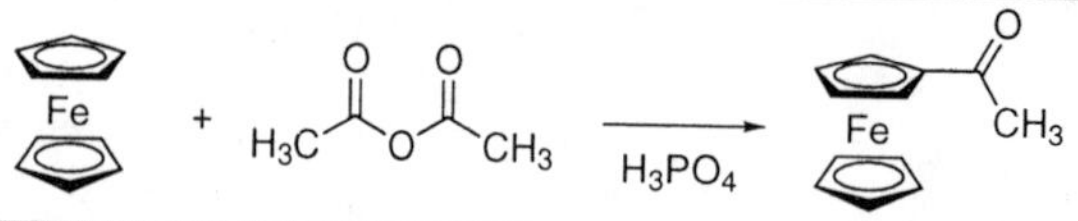

Concepts/techniques:

Electrophilic aromatic substitution; recrystallization (hot filtration, use of decolorizing carbon); thin-layer chromatography; melting point determination.

Green chemistry messages:

Safer reagents.

19. COMBINATORIAL CHEMISTRY: ANTIBIOTIC DRUG DISCOVERY (P. 231 - 238)

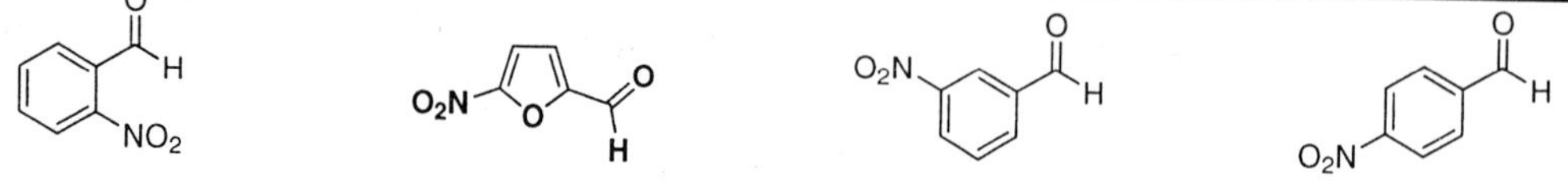

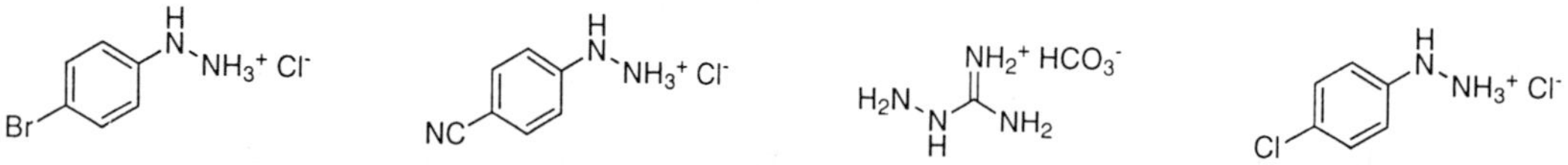

Concepts/techniques:

Combinatorial chemistry; carbonyl chemistry; antibiotics; sterile techniques; biological assays.

Green chemistry messages:

Synthetic efficiency; design of safer products; waste minimization.

Chapter 1: Introduction

This chapter provides a brief overview of the role of organic chemistry in modern life and of the development of modern organic chemical industry. Following a discussion of the hazards to health and environment resulting from the development of this industry, the general concept of "green" organic chemistry is introduced.

1.1 The Many Benefits of Organic Chemistry – The Impact of Organic Chemistry on Our Lives

As you have undoubtedly learned through your previous courses in chemistry, chemical compounds are important in virtually all aspects of modern life. Organic chemistry and organic chemicals play profound roles in our lives, representing the compounds responsible for life itself (e.g., DNA, proteins, and carbohydrates) and ranging from the clothes we wear to the pharmaceutical products we rely on to treat illnesses. Often, organic chemicals may play less prominent but no less important roles, not ending up in consumer products, but essential in the manufacturing processes for these products. Representative examples of the enormous range of roles and uses of organic compounds are listed below.

Biochemistry – proteins, enzymes, nucleic acids (DNA and RNA), hormones, membranes

Medicine – drugs, biocompatible implants

Electronics – liquid crystals, light emitting diodes (LEDs), insulators, photoresists

Polymers/plastics – polyethylene, polypropylene, polystyrene, polycarbonates

Paints and coatings – colorants, cross-linkers

Cosmetics, sunscreens

Insect repellents and pheromones

Dyes, photographic film, printing

Foods and additives

Fuels, lubricants – natural and synthetic

High energy content materials – explosives, propellants

Textiles – dyes and fibers

Refrigerants – air conditioning, home and commercial cooling

And many others…New chemicals and materials are being discovered/created every day

As a society we have come to depend on the products of organic chemistry. In many cases we take these products for granted. But what is the cost we pay for these products, which in some cases are necessities but in others merely conveniences?

Worldwide, chemical manufacture represents an industry with an annual value of approximately $1,500,000,000,000, (global chemicals industry sales in 1998) [1], corresponding to the production of countless millions of tons of plastics, fuels, pharmaceutical and agricultural agents, and numerous other bulk (large volume) and commodity (specialty) chemicals. While these numbers refer to overall chemical sales, organic chemicals make up a significant fraction of these overall numbers. The magnitude of this industry is difficult to fathom. As but one example, a single chemical facility recently brought on-line (MCC PTA India Corporation Private Limited, Haldia, West Bengal Province, India) has the capacity to produce 350,000 tons per year of terephthalic acid, used in the manufacture of polyester derivatives such as polyester fiber, poly(ethylene terephthalate) (PET), and other products.

Before Wöhler's discovery in 1828 that urea, the prototypical organic compound, could be prepared from inorganic precursors, it was generally felt that the manufacture of organic compounds, "the chemicals of life," could only be effected by life processes, not in the laboratory. In this light, the scale of modern organic chemical industry is all the more remarkable – even the existence of such an industry, let alone its magnitude, would have been essentially inconceivable in Wöhler's time. Nonetheless, while organic chemistry is a relatively new field, it has clearly made a remarkable impact on our lives over the past century and a half, and it will continue to do so in the future.

The development of the organic dye industry in mid-19th century England and Europe [2], astounding in its success given the still-developing state of affairs regarding organic chemical structure and

---

1.   "Environmental Outlook for the Chemicals Industry," Organisation for Economic Co-Operation and Development Report (2001).
2.   A fascinating account is provided by Simon Garfield in "Mauve: How One Man Invented a Color That Changed the World;" Norton: New York, 2001.

bonding at the time, signaled the arrival of the organic chemical industry. Given the excitement with which the artificial production of previously rare and costly dyestuffs was received at the time, perhaps the relative lack of attention to or concern for environmental issues associated with chemical manufacture was understandable. Chemical wastes were vented to the atmosphere or dumped into rivers, lakes, or landfills. The world, after all, was a very big place, the scale of chemical production was relatively small, the value of the chemicals being produced was very high, and the hazards of chemicals were not greatly appreciated. This spirit is captured even in the published literature of organic chemistry from these earlier days.

An early chemist exploring the class of compounds known as isocyanides, containing the R-$N\equiv C$ functional group, moved his research outdoors when the overpoweringly disgusting odor of the compounds drove him there, then ceased work with them entirely when the complaints of neighbors became too loud [3].

The discoverer of mustard gas ($ClCH_2CH_2SCH_2CH_2Cl$), a potent blistering agent used as a warfare agent in World War I, reported the terrible effects of his newly prepared compound on the nasal membranes when sniffed. In keeping with sound organic chemical practice of the times, he then tasted his compound! (If you are curious about the outcome, it caused a violent headache, dutifully reported in the manuscript reporting the isolation and analysis of mustard gas[4].)

1.2   The Need for Sustainable Chemistry

As both the chemical industry and our understanding of the potential health and environmental impacts of chemicals have grown, the inappropriateness of such naïve approaches to chemical hazards has become increasingly apparent. The unchecked emissions of hazardous materials by the chemical plants of the late 1800's to early 1900's are clearly no longer tolerable.

3.   W. Lietke, *Justus Liebigs Annalen de Chemie* **1859**, *112*, 316 (quoted in J. A. Green, II & P. T. Hoffman, "Isonitrile Chemistry," I. Ugi, Ed.; Academic Press: New York, 1971, p. 1).

4.   F. Guthrie, *J. Chemical Society* **1860**, *12*, 109 (quoted in E. E. Reid, "Organic Chemistry of Bivalent Sulfur," Vol. 2; Chemical Publishing Co., Inc.: New York, 1960, pp. 238ff).

In response to a number of issues, including health concerns, concern for the environment, compliance with legal regulations, and simple economics (waste may be viewed as lost profits), the chemical industry is undergoing an increasingly dramatic change. (These issues are discussed in greater detail below.) Much, however, remains to be accomplished. For example, in 1994, an estimated 2.26 billion pounds of hazardous substances were released into the environment – that is over 6 million pounds per day, or more than two tons per minute!

These numbers come from the "Toxics Release Inventory" (TRI) that tracks reported releases of the most toxic materials from TRI-reporting facilities [5]. In 1994, the TRI, established under the Emergency Planning and Community Right-to-Know Act of 1986, tracked roughly 300 compounds, selected in compliance with TRI statutes, which call for listing of compounds displaying acute human health risks, cancer or chronic (non-cancer) human health effects, and/or environmental effects. The TRI does not document releases from "minor" polluters (using less than 10,000 pounds of any listed chemical per year) or release of other hazardous materials that are not on the TRI list. Thus, the total amount of chemicals released per year is higher still. As of 2001, there were nearly 650 compounds on the TRI list, and releases from the "original industries" (i.e., those reporting in 1994) were estimated at 2.33 billion pounds. (Given the dramatic increase in number of listed compounds, the relatively minor increase in TRI releases over the five year period 1994-1999 might be considered to be encouraging.) New industries have been brought under TRI reporting requirements since 1994, and the 2001 estimates, including these new industries, rise to 7.77 billion pounds per year! In case these numbers are not sufficiently frightening, remember that these numbers reflect only releases from American chemical industries.

It is our hope that, through your studies of the practice of environmentally benign ("green") organic chemistry, you will find yourself prepared to work to improve this dire situation for the next generation.

---

5. Information about the TRI and the most recent data may be obtained from the United States Environmental Protection Agency through the Internet (http://www.epa.gov/tri/).

1.3 Green Chemistry: New Approaches to Important Problems

When a reaction or process presents a chemical hazard, one can attempt to minimize human and environmental exposure to the hazard, and one can seek to reduce or eliminate the hazard associated with the process. Traditionally, emphasis has been placed on the former – developing techniques or facilities allowing one to minimize personal exposure and minimizing release to the environment. In contrast to the early days of organic chemical research discussed above, current laboratory practice emphasizes common sense in the handling of chemicals. Now routine is the use of gloves or other protective gear to avoid contact with and absorption of chemicals through the skin, avoidance of ingestion, and use of well-ventilated fume hoods and laboratories to prevent exposure to volatile hazardous chemicals. However, despite our best efforts, some level of exposure is difficult to avoid, and occasional accidents can lead to significant exposure. Further, while dispersal of volatile compounds into the atmosphere through fume hoods protects the individual working with the compound, it does not represent a sound long-term solution or, for that matter, even a particularly good solution for certain immediate exposure problems. (Woe to the maintenance worker, or the pigeons, on the roof of a chemistry building when a researcher chooses to vent highly toxic hydrogen cyanide from a reaction vessel into the fume hood.)

Green Chemistry has been defined [6] as "the utilization of a set of principles that reduces or eliminates the use or generation of hazardous substances in the design, manufacture, and application of chemical products." Rather than focusing on reduction of chemical exposure and environmental contamination through "environmental controls" such as gloves and fume hoods, green chemistry seeks to reduce or eliminate hazards through the replacement of hazardous substances with less- or non-hazardous substitutes. Through such approaches, personal safety is assured even in cases of accidental exposure. In addition, green chemistry addresses long-term environmental issues by avoiding the release of hazardous materials and through the minimization of hazardous waste generation.

In order to address rationally the issue of minimization of chemical hazards, whether through the traditional approach of environmental controls (avoidance of exposure) or through the development of green chemical practices and procedures, one must first be able to identify chemical hazards. The next chapter presents and discusses some of the issues associated with the identification and quantification of chemical hazards.

Summary

Organic chemistry and organic chemicals play key roles in virtually all aspects of our lives, ranging from the fundamental biochemical processes responsible for life itself to the clothing we wear, the food we eat, and the medicines we take. As the scale of chemical industry has increased and as we have become more aware of the potential health and environmental hazards of chemicals, it has become increasingly more important to deal with chemicals not just safely, but wisely. The "Toxics Release Inventory" (TRI) provides for tracking of the release into the environment of roughly 650 compounds recognized as being particularly hazardous. Green chemistry seeks to reduce the hazards associated with chemical processes not just by preventing exposure or release, but by reducing the intrinsic hazards of the chemicals employed.

---

6.   P. T. Anastas and J. C. Warner, "Green Chemistry: Theory and Practice;" Oxford University Press: Oxford, 1998.

Chapter 2: Identification and Evaluation of Chemical Hazards

This chapter introduces the concepts of chemical toxicity and hazards. General types of chemical hazards are presented, and measures for the quantitative evaluation of these hazards are discussed. Several of the most important classes of hazardous compounds are surveyed.

2.1   Chemical Hazards

"Poison is in everything, and no thing is without poison. The dose makes it either a poison or a remedy."

Paracelsus, 1564

Paracelsus, recognized as the father of modern medicine, observed that certain substances could be either harmful or beneficial to human health and further noted that, often, the harm or benefit provided by such substances was dependent upon their dosage. For example, table salt (NaCl) and aspirin (acetylsalicylic acid) can be lethal if taken in sufficiently large dosage, and even water can be fatal, with ingestion of large quantities leading to seizures or death. We now know much more about the various hazards associated with chemicals and have developed ways to quantify, and compare, the relative hazards of these chemicals. As discussed in detail in this chapter, we categorize materials, based upon the ways in which they cause harm, into groups, including carcinogens, cancer-suspect agents, mutagens, teratogens, tumor promoters, corrosives, and lachrymators. Methods used to quantify hazards include animal (usually mammal) studies, bacterial assays, and human epidemiological studies. By measuring and categorizing hazards, we can begin to assess the risks associated with exposure to various chemicals and classes of chemicals. In the following sections, we will explore some of the most important chemical hazards and the ways in which these hazards are quantified for use in the assessment of risk. (Risk assessment itself is discussed in Chapter 4.)

Before we continue this discussion, it is important to recognize that not all chemicals are hazardous. Indeed, the natural substances necessary for life itself are chemicals. Thus, we must not *assume* that all chemicals are hazardous; instead, we must learn about their hazards and treat them appropriately. We should, at the same time, not make the mistake of assuming that all things natural are without risk. As you will see in the examples below, many materials that we think of as "natural" represent very serious hazards.

Acute chemical hazards are those which lead to immediate health consequences, while the health implications of **chronic chemical hazards** are the result of cumulative exposure. Well-known examples of acute hazards include hydrogen cyanide, used for the application of capital punishment in the "gas chamber," and "nerve gases" such as Sarin ("GB") and VX, stockpiled for potential use in military operations. A single exposure to such compounds can lead to immediate and dramatic health consequences. In contrast, a single exposure to lead or mercury compounds may or may not result in observable health effects. It is generally the cumulative effects of longer-term (chronic) exposure to such compounds that lead to serious problems.

Molecular structures of typical nerve gases

2.2   Quantitative Evaluation of Chemical Hazards

Almonds contain measurable amounts of toxic cyanide, automobile gasoline contains cancer-causing benzene, cancer chemotherapy agents can themselves cause cancer – every day, we make conscious or subconscious choices about chemical exposure. The question, ultimately, is "How hazardous is this compound?" In other words, how do we quantify chemical hazards? In order to evaluate the possible risk of exposure to a compound (for example through physical contact, inhalation or ingestion), one must be able to evaluate hazard.

Toxicity. A commonly used measure for the quantitative evaluation of acute toxicity is the "lethal dose 50," or LD_{50} measure. This measure, usually provided in the units of grams of substance per kilogram of body mass, represents the dosage of a compound that results in the death of 50% of an animal population (usually rats) within one week of administration. Table 2.1 provides a representative listing of a variety of substances and their corresponding LD_{50} values. Interestingly, naturally occurring toxins are frequently far more toxic than either simple inorganic compounds or manmade toxins. For example, gram for gram, botulinium toxin is 500,000,000 times more toxic than sodium cyanide and 10,000,000 more toxic than Sarin nerve gas.

Table 2.1. LD_{50} values for various chemicals

Substance	LD_{50} (grams/kilogram)
Water	180
Sugar	35
Sodium chloride	3.75
Caffeine	0.13
Sodium cyanide	0.015
Arsenic trioxide	0.015
Aflatoxin (moldy legumes and grains)	0.010
Sarin (nerve gas)	4×10^{-4}
Tetanus toxin A	5×10^{-9}
Botulinium toxin	3×10^{-11}

Part of the reason for the enormous range of LD_{50} values for compounds (i.e., for the tremendous range in toxicity) is that there are many mechanisms of toxicity. Some toxins may cause specific organ or tissue damage (e.g., to the liver or bone marrow), others may interfere with the normal workings of enzymes, and still others may change the distribution of ions (e.g., K^+, Ca^{2+}) in the interior and exterior of a cell. For these and myriad other toxicity mechanisms, the activity of a particular toxin can be dramatically dependent on both the nature of the toxin and the nature of the cellular or biochemical target for its activity.

Traditionally, LD_{50} values have formed the basis for guidelines (such as permissible exposure limits – "PEL") regarding acceptable exposure to substances. In green chemistry, as we discuss in later chapters, the emphasis is quite different – rather than focusing on limiting exposure, we aim to find less toxic alternatives that will serve the same purpose.

When using LD_{50} values or any other indicators of chemical hazard, it is important to recognize that two hazardous substances acting together can either accentuate or nullify the each other's harmful effects. For example, whereas both mercury and selenium compounds can be quite toxic, the toxicity of one appears to be *reduced* by concurrent exposure to the other. The mechanism of toxicity reduction in this case appears to involve the precipitation of highly insoluble mercury-selenium compounds (apparently closely associated with sulfur-containing amino acid groups in proteins), effectively removing both mercury and selenium from solution and reducing their undesirable properties.

Clearly, the issue of quantitative assessment of chemical hazards is complex, and in many cases quantification is an inexact science that can be used only as a rough guide in hazard evaluation. LD_{50} values address only acute toxicity. No analogous parameters are available for the quantification of chronic toxicity, which is generally assayed by exploring the relationship between chronic human exposure and disease (epidemiological studies – see below).

2.3   Classes of Hazardous Substances

Hazardous chemicals may be categorized in classes according to the nature of their deleterious effects in humans. The following discussions introduce several of the more important and more commonly encountered classes of chemical hazards, distinguished either by their mechanisms of action upon chronic exposure (e.g., carcinogenesis) or by the immediate results of acute exposure (e.g., lachrymation).

Carcinogens. Epidemiological studies strive to determine the environmental factors that lead to diseases by seeking unambiguous links between exposure to particular substances and incidences of disease. If substantial evidence from epidemiological studies suggests that a substance actually causes cancer in animals and/or humans, the substance is classified as a ***carcinogen***. This is hardly a new concept – as long ago as 1775, it was noted that chimney sweeps, with prolonged exposure to soot, had elevated incidences of cancer. One of the active cancer-causing principles found in soot, benzo[a]pyrene, is also found in cigarette smoke, charred meat, and diesel exhaust. Other known human carcinogens include 1-aminonaphthalene, 2-aminonaphthalene, benzidine, N-nitrosodimethylamine, and 1,2-dibromo-3-chloropropane.

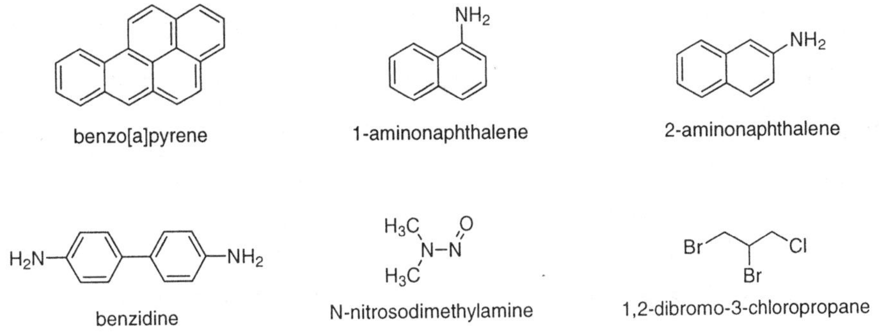

Representative carcinogens

The carcinogenic activity of so-called polycyclic aromatic hydrocarbons (PAHs) such as benzo[a]pyrene is interesting because it is the action of our own biochemical defense systems that converts the PAH into a carcinogen. When a PAH enters the body (perhaps through inhalation of diesel exhaust fumes), it is recognized as a foreign element, and the body's waste elimination processes go to work. The liver contains an enzyme that serves as a potent oxidizing agent, capable of introducing oxygen-containing functionality (e.g., alcohol, ketone, epoxide) into even simple hydrocarbons. By so doing, the liver converts such compounds to water-soluble derivatives, which can then be excreted in the urine.

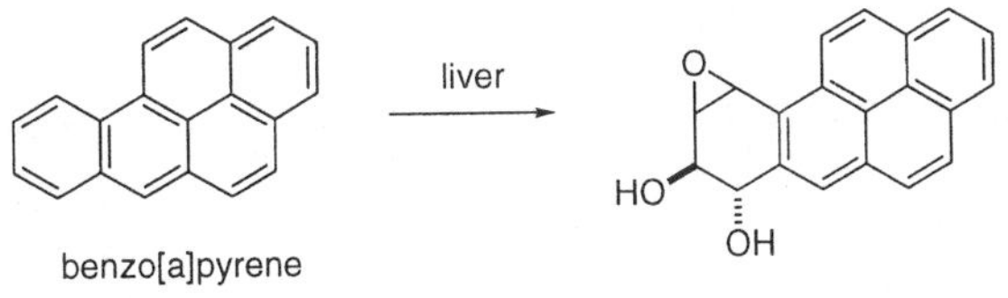

Conversion of benzo[a]pyrene to a highly carcinogenic diol epoxide

Unfortunately, in the case of the PAHs, the oxidized derivatives can act as potent carcinogens. Perhaps it would be better if the liver simply left these insoluble compounds alone rather than working so desperately to excrete them (although the long-term health consequences of buildup of such insoluble compounds remain unknown).

Cancer-suspect agents. In the absence of epidemiological studies, a compound may be classified as a *cancer-suspect agent* based on its *structural or chemical similarities to known carcinogens*. Polycyclic

29

aromatic hydrocarbons, particularly those with a "bay region" (a portion of the molecule displaying a concave shape, as illustrated in the following figure), are high on the list.

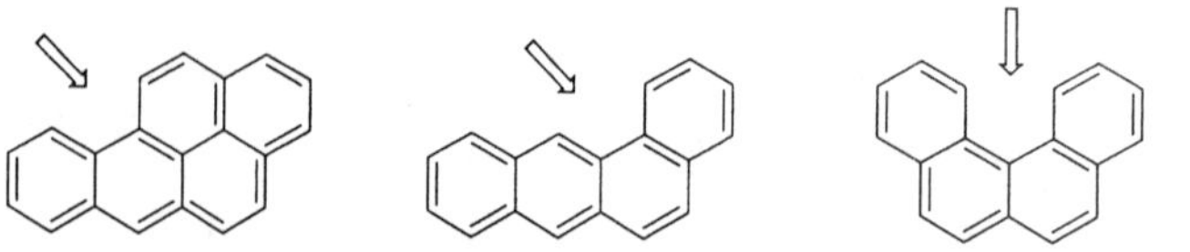

benzo[a]pyrene   benz[a]anthracene   benzo[c]phenanthrene

Polycyclic aromatic hydrocarbons with "bay region" indicated by arrow

Functionalized analogs of polycyclic aromatic hydrocarbons (e.g., 1-nitropyrene, 6-nitrochrysene), N-nitrosamines and powerful alkylating agents (e.g., 1,2,3-trichloropropane) are also frequently considered cancer-suspect agents.

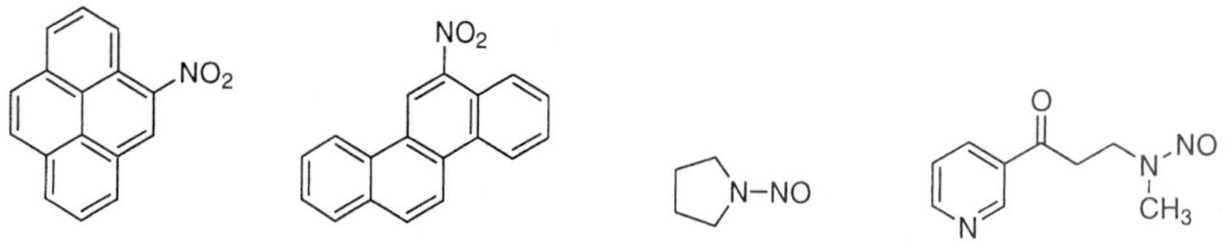

1-nitropyrene   6-nitrochrysene   1-nitrosopyrrolidine
(fried bacon)   4-(methylnitrosoamino)-
1-(3-pyridyl)-1-butanone
(tobacco smoke)

Other cancer suspect agents

Mutagens. Compounds that cause mutations in the DNA of bacteria, as revealed by the so-called Ames test [7], are referred to as ***mutagens***. Since DNA damage can, in principle, lead to cancer, mutagenicity can be an alert to potential carcinogenicity. The elementary reaction chemistry of DNA is the same regardless of the origin of the DNA (whether it be human, plant, or bacterial). Thus, a compound that causes mutations in bacteria (i.e., that causes DNA damage in bacteria) might be inferred to have the potential to cause DNA damage and/or cancer in humans. This thought process is at the heart of the Ames test. The test is simple to carry out – bacteria typically have life cycles ranging from a few hours to a few days, so that mutations can quickly become apparent, and they are far easier to cultivate and analyze than laboratory mammals. However, the Ames test is highly controversial, with detractors noting that the ability to cause mutation of bacterial DNA does not *necessarily* predict cancer in humans. There are many natural DNA repair mechanisms that can prevent a mutation from leading to cancer. Because these repair mechanisms differ from organism to organism, mutations found in bacteria might not translate into cancer in humans. In addition, if the actual cause for mutation and/or carcinogenesis is not the compound being tested, but rather a compound derived metabolically from it, the Ames test could either inappropriately label a safe compound as a carcinogen or suggest a compound to be safe when it is not. Representative mutagens include 2-aminopurine, 5-bromouracil, hydroxylamine, and ethyl methanesulfonate.

2-aminopurine   5-bromouracil   hydroxylamine   ethyl methanesulfonate

Typical mutagens

Teratogens. Substances that cause developmental malformations (birth defects) are known as *teratogens*. The list of known or suspected teratogens is remarkably long, and includes a diverse array of compounds, ranging from simple inorganic materials (e.g., cadmium and its salts, carbon monoxide) to common organic solvents (e.g., benzene, chloroform, N,N-dimethylformamide, toluene – found in airplane glue) to complex molecules of synthetic or biological origin (e.g., librium, testosterone). One of the best-known and most notorious teratogens is the (*R*)-enantiomer of the once commonly prescribed drug, thalidomide. While the (*S*)-enantiomer exhibits excellent activity in the relief of morning sickness in pregnant women, the (*R*)-enantiomer is a potent teratogen. This drug is currently undergoing something of a renaissance, displaying promising activity for the treatment of leprosy, but due to the extreme teratogenicity of the (*R*)-enantiomer, which is readily formed in equilibrium from the (*S*)-enantiomer under physiological conditions, its use must be carefully controlled.

---

7.   R. Devoret, "Bacterial Tests for Potential Carcinogens," *Scientific American* **1979**, *241(2)*, 40-49.

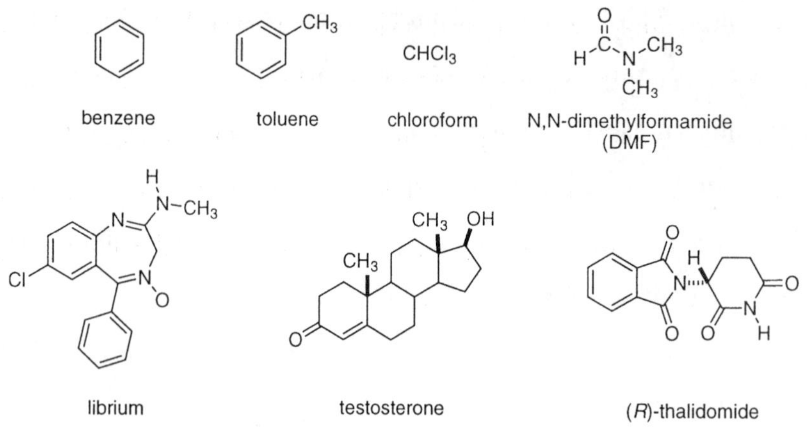

benzene | toluene | chloroform | N,N-dimethylformamide (DMF)

librium | testosterone | (*R*)-thalidomide

Typical teratogens

Tumor promoters. A variety of compounds have been discovered that are not themselves carcinogenic, but which enhance the carcinogenicity of other compounds. These compounds, known as **tumor promoters** and helping form the basis for interesting new research approaches to understanding how cancer arises, should be treated with due respect. One example of such activity is found in the phorbol esters and related compounds, naturally occurring complex organic compounds found in the sap of plants of the *Euphorbiaceae* family.

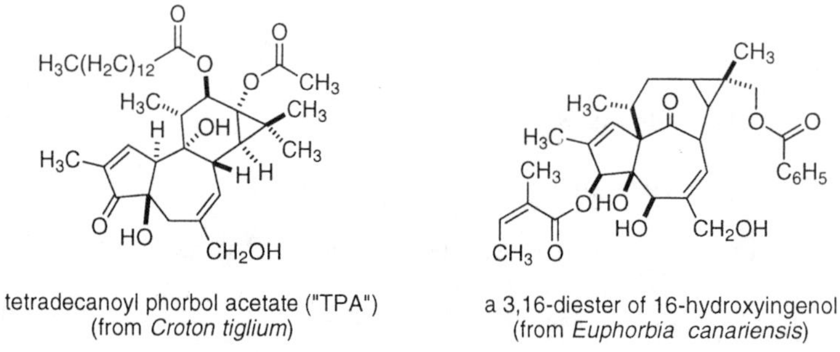

tetradecanoyl phorbol acetate ("TPA")
(from *Croton tiglium*)

a 3,16-diester of 16-hydroxyingenol
(from *Euphorbia canariensis*)

Typical phorbol esters

Corrosives. This well-known class of hazardous substances includes strong acids and bases as well as compounds that form acids or bases upon decomposition. Such compounds can cause severe skin and tissue damage.

Lachrymators. Some compounds, including bromoacetone, benzyl bromide, chloroacetophenone, and ethyl iodoacetate, are remarkably irritating to mucous membranes (eyes, nasal passages), and exposure to their vapors causes severe eye watering (lachrymation). Both solid and liquid compounds can display sufficiently high vapor pressures to act as **lachrymators**, and such compounds form the basis of "tear gas" agents used to effect the dispersal of unruly crowds or the temporary incapacitation of attackers.

| bromoacetone | benzyl bromide | chloroacetophenone | ethyl iodoacetate |

Typical lachrymators

Unstable compounds. Certain classes of compounds are shock sensitive and can explosively decompose. Whereas some explosives require a primer to set them off, others, known as primary explosives, do not and are thus particularly dangerous. Organic peroxides and inorganic perchlorate salts are notoriously explosive, and polynitrated organic compounds (e.g., nitroglycerin, hexanitrohexaazaisowurtzitane) are also well-known for their explosive properties.

| cumene hydroperoxide | dibenzoyl peroxide | nitroglycerin | hexanitrohexaaza-isowurtzitane |

Typical unstable compounds

As the preceding discussion of chemical hazards should make clear, there are a number of different ways in which chemicals can cause adverse health effects. The magnitude of the hazard presented by an individual chemical may be estimated by measuring the chemical's effects in animal, bacterial, or human epidemiological studies. In some cases, including that of cancer-suspect agents, we have seen that it has been possible to correlate the structure of compounds with their *potential* risks. This type of correlation is generally referred to as a "structure/activity relationship." As scientists learn more about how chemical hazards depend upon specific chemical structure, we may ultimately be able to predict fairly accurately the hazards of most substances.

Summary

The human health hazards presented by a substance are determined by not only the intrinsic properties of the substance, but also by the quantity (dosage) of the substance to which one is exposed. Some substances display acute toxicity, apparent after a single exposure, while others represent chronic exposure hazards, with health problems resulting only from repeated exposure over a period of time. The intrinsic acute toxicity of a substance may be quantified in the form of LD_{50} values, representing the dosage that proves lethal to 50% of a test animal population. Hazardous chemicals exert their toxicity in a variety of ways, related to their physical and biochemical properties, and may be grouped into various classes, including carcinogens (both known and suspected), mutagens, teratogens, tumor promoters, lachrymators, and unstable (explosive) compounds. While the members of each of these classes of hazardous substances have for the most part been recognized on the basis of their observed human or animal health effects, scientists are improving their ability to predict chemical hazards on the basis of molecular structure.

Chapter 3: Chemical Exposure and Environmental Contamination

This chapter describes the mechanisms leading to release of chemical substances into the environment and the environmental impacts of the released chemicals. By understanding these mechanisms and environmental impacts, we can learn to choose chemicals that are less likely to be released and have less impact on the Earth's atmosphere, water and soil.

3.1   Health Risks and Environmental Impact Require Exposure

With all the potential hazards discussed in the preceding chapter, it may seem a wonder that anyone either wants to become a chemist or lives long enough to do so! Fortunately (and important to note even though seemingly obvious), for a chemical to pose a human risk, one must be exposed to it. Similarly, for a compound to harm the environment, it must be released to the environment. Unfortunately, there are many potential ways to be exposed to hazardous substances, some accidental, others not. It is thus important to consider issues related to chemical exposure in order to learn both how to protect ourselves and how to find ways to prevent exposing others to hazards, whether through environmental controls, green chemistry, or both.

Although the list of known or potentially hazardous compounds may seem very long, many organic compounds appear to have no significant health or environmental impacts. In some cases, this is because they simply do not have any harmful properties. In other cases, their lack of significant impacts is due to the fact that their physical properties do not offer facile mechanisms for exposure or emission. Two physical properties of compounds are often very significant in determining their potential for exposure or emission – volatility (ease of evaporation) and solubility (particularly in water, given the role water plays as the essential solvent in living organisms and as a transport agent in the environment). If a compound has a negligible vapor pressure, then the risk of exposure to the compound through inhalation, obviously, is vanishingly small. Only direct contact with or ingestion of the compound can present a risk in such cases. Similarly, if a compound is absolutely insoluble in water, the chance of it mobilizing within the body upon exposure or in the environment upon release is very small. Perhaps not surprisingly, there are notorious exceptions to these general statements about volatility and solubility. For example, complex polycyclic aromatic hydrocarbons like those described in our discussions of carcinogenicity in Chapter 2 have very low vapor pressures and are virtually insoluble in water, yet are recognized as potent carcinogens.

3.2   Volatile Organic Compounds (VOCs)

Organic compounds with high volatility are used as solvents, plasticizers (e.g., "new car" smell) and cleaners, as well as in many other applications. While it is not surprising that volatile organic compounds (VOCs) comprise a major source of our daily chemical exposure, the primary sources of our exposure to VOCs may be. Consumer products (e.g., aerosol sprays and room air deodorants), building materials (e.g., paints and adhesives), and personal activities (e.g., smoking and driving) result in the vast majority of our daily exposure to VOCs. Even taking a shower contributes to our exposure to VOCs, through the volatilization of traces of chloroform found in chlorinated tap water. The more visible sources of exposure such as chemical plants, petroleum refineries, and hazardous waste sites, in contrast, account for less than 25% of total human exposure. VOCs contribute to health problems when used in areas without adequate ventilation, may contribute to stratospheric ozone depletion and global warming, and can result in water pollution when rain dissolves these materials and deposits them in our water supplies.

3.3   Emissions from Human Activities

In order for a substance to have an impact on our environment, it must first be released into the air, water or land through accidental, unintentional or deliberate action. Although accidental chemical releases frequently receive a great deal of attention, they are in fact comparatively uncommon. More typically, chemical emissions occur in the form of either unintentional or intentional releases. Unintentional releases can arise when the nature of a substance (e.g., volatility or water solubility) makes control of its emission difficult. Intentional releases are the result of legally permissible (or in some cases illegal) discharge of chemicals into the air, water, or land. Some of these emissions are lessened by environmental controls such as smokestack scrubbers and water treatment facilities, but the volume of chemicals being introduced to our environment remains staggering. Following brief discussions of chemical emissions into the air, water, and land, we will see how the principles of green chemistry can be applied to minimize or eliminate chemical release and environmental impacts.

Air pollution. Whereas solids (particulates) can and do represent a significant component of air pollution, the primary property of a substance that contributes to air emission is its volatility. Air pollution results from the generation and release of volatile (gaseous or high vapor pressure liquid or solid) materials during chemical transformations (including combustion) and from the evaporation of volatile organic solvents and compounds. Introduction of chemicals into the atmosphere leads to a number of significant concerns. Atmospheric pollutants can lead to either acute or chronic health hazards – for example, benzene, a component of automobile gasoline, is volatile and a known carcinogen. In addition, the chemical reactivity of compounds in the atmosphere can lead to progressive and deleterious impacts, including photochemical smog, ozone depletion, and global warming.

Photochemical smog. Smog results from combustion of fossil fuels like gasoline and coal. The combustion process releases a variety of pollutants, including hydrocarbons (unburned fuel) and nitric oxide (NO) into the troposphere (the lowest level of our atmosphere). Sunlight induces chemical chain reactions leading to the conversion of NO to nitrogen dioxide (NO_2), which contributes a brown color to the atmosphere. Atmospheric reactions of NO_2 in turn lead to the formation of ozone and a variety of oxidized organic compounds, including various aldehydes and ketones, as well as peroxyacetyl nitrate (PAN). These compounds are often irritating to the eyes and respiratory tissues.

peroxyacetyl nitrate (PAN)

Molecular structure of peroxyacetyl nitrate

As discussed below, whereas ozone in our upper atmosphere is essential to protect us from damaging ultraviolet radiation, ozone in the troposphere is a hazard. Ozone levels in the troposphere (roughly 30 parts per billion) are now approximately double those found in unpolluted air, leading to significant health, crop, and structural damage.

Ozone depletion. The upper atmosphere (stratosphere) contains a layer of ozone, formed from the action of ultraviolet light from the sun on molecular oxygen. This ozone layer is widely recognized as vital to life on Earth, as it prevents the most damaging higher energy ultraviolet light from the sun from reaching the Earth's surface. In 1985, a recurring "hole," representing a significant decrease in the amount of ozone in the ozone layer, was detected over Antarctica. While the most prominent "hole" is localized over Antarctica, there appears to be a global decrease in the amount of ozone present in the ozone layer. Studies have strongly implicated chlorofluorocarbons (CFCs) as the primary culprits. CFCs persist long enough to diffuse to the stratosphere, where more intense solar radiation induces their decomposition. The initial products of their decomposition lead to complex cascades of chain reactions, one outcome of which is the net destruction (depletion) of ozone. For example, photochemical reactions of CFCs lead to the formation of chlorine atoms, which react with ozone to generate chlorine oxide (ClO) and molecular oxygen. Chlorine oxide, in turn, reacts with ozone to generate molecular oxygen and regenerate chlorine. The chlorine atom, then, acts catalytically to destroy ozone, and it has been estimated that a single chlorine atom can destroy as many as 10,000 molecules of ozone.

$$CFC \xrightarrow{\text{ sunlight }} Cl^{\bullet}$$

$$2\,O_2 \quad O_3 \qquad Cl^{\bullet} \qquad O_3$$

$$O_3 \qquad ClO \qquad O_2$$

Ozone destruction by photochemically generated chlorine atoms

Chlorine oxide also effects the conversion of nitric oxide to nitrogen dioxide (discussed above) via the formation of ClONO. Such compounds act as "reservoirs" for atmospheric chlorine, holding it in an unreactive form but regenerating it upon the action of sunlight.

The story of CFCs and ozone depletion provides an important lesson regarding the unintended consequences of our actions as scientists and reminds us that it is important to constantly reexamine our hypotheses and conclusions. CFCs were invented by chemists to replace the dangerous materials (such as pure ammonia) that were previously used as refrigerants. They were thought to be excellent replacements because they are virtually inert under ordinary laboratory conditions (i.e., they are quite unreactive). Thus, they posed no apparent health risks and, in addition, appeared to be harmless in the environment. However, because of their chemical stability, the CFCs accumulate in the environment and, as we now know, do indeed have significant detrimental effects.

Global warming. Numerous human activities, the foremost of which is the combustion of fossil fuels, generate carbon dioxide. As such activities have increased, so have atmospheric CO_2 concentrations. Carbon dioxide, as well as a number of other gases released either intentionally or inadvertently (including methane and CFCs), transmit warming solar radiation to the Earth's surface while absorbing infrared radiation. Since emission of infrared radiation to outer space represents a significant mechanism by which the Earth cools, infrared absorption within the atmosphere might plausibly be expected to lead to an increase in the temperature at the Earth's surface. Some estimates suggest the temperature could increase by several degrees Celsius over the next several decades if the release of these so-called "greenhouse" gases is not curtailed. Such a warming could lead to remarkable agricultural consequences and, through the partial melting of the polar ice caps, to coastal flooding. The details of such global warming are still under active research and discussion, and this is an area of great debate, both scientific and political.

Water pollution. Water pollution results not only from the accidental or intentional release of contaminated water, but also from the action of water ("leaching") on air pollutants (e.g. acid rain) and on solid waste disposal or storage sites. The major property of a substance that increases the likelihood of its release into water is thus its solubility in water. Typically the more water-soluble a compound is, the more likely it will be carried into the environment along with wastewater or will be "leached" (dissolved and washed away) from solid waste sites. Sources of water pollution are ubiquitous, and include even the chlorination processes used to purify water for human consumption – through the chlorination process, some organic pollutants are converted to undesirable chlorinated compounds such as chloroform. Water quality can also be adversely affected by large changes in temperature or pH as well as by chemical contamination.

Soil pollution from solid and liquid waste. Soil pollution results from the accidental or intentional release of solid and liquid wastes. Such wastes include substances used in chemical processing that cannot be easily reused. In earlier days, solid and liquid wastes were frequently simply buried in landfills, leading to a number of notorious environmental disasters.

> *Love Canal, in Niagara Falls, New York, was used as the burial site for an estimated 21,000 tons of chemical waste from the 1920's through the early 1950's [8]. Hundreds of chemicals, including a number of recognized or suspected carcinogens, were dumped. After the site was capped with soil, a school and, eventually, some eight hundred homes and 240 low-income apartments were built on or adjacent to the dumpsite. Heavy rains in the late 1970's led to significant leaching, and foul puddles of waste formed in and around homes and the schoolyard. Contact with these wastes resulted in chemical burns, and longer-term studies have suggested higher than normal rates of miscarriage and birth defects. Following action by New York Governor Hugh Carey and President Jimmy Carter, upwards of two hundred homes were purchased by the state of New York and evacuated.*

Recognition of the unacceptable nature of this practice has led to more appropriate disposal practice for wastes. Solid wastes that present no environmental or health hazards are treated essentially the same as domestic trash and buried in landfills. More hazardous substances are often treated to decrease their solubility (in an effort to reduce the possibility of leaching) and then buried in landfills. Solid waste is also disposed of by burning (defined as combustion under conditions allowing recovery of energy from the combustion process) or by incineration (combustion without energy recovery). Liquid waste, when not recyclable, is generally either burned or incinerated. While complete combustion, under ideal circumstances, could convert organic compounds to carbon dioxide (a greenhouse gas!) and water, incomplete combustion and the presence of other elements, particularly halogens, in waste streams can lead to the emission of hazardous combustion products. For example, the combustion of a

---

8.   E. C. Beck, "The Love Canal Tragedy," *EPA Journal*, January, 1979.
      (http://www.epa.gov/history/topics/lovecanal/01.htm).

wide variety of compounds commonly found in solid municipal waste leads to the formation of polychlorinated benzodioxins (referred to generically as "dioxins"), including 2,3,7,8-tetrachlorodibenzo-*para*-dioxin (TCDD). TCDD is a potent human carcinogen also known to cause severe reproductive and developmental problems and immune system damage.

2,3,7,8-tetrachlorodibenzo-*para*-dioxin
(TCDD)

Molecular structure of TCDD, a common byproduct of combustion

3.4   Natural Sources of Environmental Exposure to Hazardous Substances

A variety of natural processes can and do lead to the introduction of substantial amounts of hazardous compounds to the environment. Volcanic eruptions, for example, release enormous quantities of carbon dioxide, sulfur dioxide, hydrogen chloride, hydrogen fluoride, and particulate matter into the atmosphere. The 1991 eruption of Mount Pinatubo, in the Philippines, injected a 20,000,000-ton cloud of sulfur dioxide into the stratosphere. The resulting sulfate aerosol cooled the Earth's surface by as much as 1.3 °F and led to the acceleration of ozone destruction, resulting in the lowest yet recorded stratospheric ozone levels [9].

The earlier eruption of Mount St. Helens, in southern Washington State, released roughly 1,500,000 tons of sulfur dioxide. A relatively gentle eruption of the Laki crater-row in Iceland occurred over an 8-9 month period in 1783. The resulting sulfate haze destroyed most of the summer agricultural crops, leading in turn to the death of 75% of livestock and 24% of the human population, and the haze drifted across Europe during the 1783-1784 winter, which proved unusually cold [10].

9.   K. A. McGee, M P. Doukas, R. Kessler, and T. M. Gerlach, "Impacts of Volcanic Gases on Climate, the Environment, and People," U.S. Geological Survey Open-File Report 97-262, May, 1997 (http://pubs.usgs.gov/openfile/of97-262/of97-262.html).
10.   H. Sigurdsson, "Volcanic pollution and climate – the 1783 Laki eruption," American Geophysical Union, *EOS Transactions*, **1982**, *10 (August)*, 601-602.

Compounds such as methyl chloride and methyl bromide are released through natural processes in the oceans and salt marshes. Recent studies suggest as much as 40% of atmospheric methyl bromide (perhaps 14-15,000 tons) arises from these natural sources, with the remaining amounts coming from fumigation, burning of vegetation, and other sources. Forest fires are major contributors to air pollution, and many plants (e.g., pine, eucalyptus, and poplar trees) emit significant quantities of volatile organic compounds.

There is little we can do to curb natural emissions, so while it is important that we be aware of such natural sources, they are not the focus of this discussion. In addition, as large as these natural releases may seem, they are small compared to the amount of pollutants released through human activities. Thus, we focus here, and throughout this text, on those hazards over which we potentially have some control.

Now that we are aware of the types of chemical hazards and the mechanisms by which substances are released into the environment, we need to learn to assess the hazards associated with individual substances and, ultimately, develop methods for minimizing or eliminating those hazards. These topics are discussed in the following chapters.

Summary

Chemical substances are continually being released into our environment through both human actions and natural processes, and as a consequence, we are exposed to these substances. From the standpoint of controlling or preventing the release of these materials, those compounds that are volatile or water-soluble are the most problematic. Once in the environment, these hazardous compounds can cause human health problems or adversely affect other organisms, destroy property, and contribute to global climate change by interfering with the natural processes in our ecosystem.

Chapter 4: Sources of Information about Chemical Hazards

This chapter briefly discusses the evaluation of chemical hazards through the analysis of published chemical safety information and presents a listing of sources of information regarding the health risks and environmental impacts of chemicals.

4.1   Evaluation of Hazards

Central to the development of green chemical alternatives is the ability to assess knowledgeably and intelligently the health hazards and environmental impacts presented by chemicals and processes. In Chapters 2 and 3, we discussed the health and environmental hazards posed by chemicals and explored the routes by which they are released into the environment. Here we will describe some of the most important resources used to evaluate the hazards of chemicals used in the lab.

Numerous sources of information regarding the health risks and environmental impacts of chemicals are available. Perhaps the best known sources are the Materials Safety Data Sheets ("MSDS"), which are available for virtually every commercially available compound. These data sheets can be overwhelming, but they contain a wealth of information. While the specific form may vary depending on the source of the MSDS, each will present information about chemical properties, hazards, and safety precautions, divided into the following sections.

1. Product Identification
2. Composition/Information on Ingredients
3. Hazards Identification
4. First Aid Measures
5. Fire Fighting Measures
6. Accidental Release Measures
7. Handling and Storage
8. Exposure Controls/Personal Protection
9. Physical and Chemical Properties

10. Stability and Reactivity

11. Toxicological Information

12. Ecological Information

13. Disposal Considerations

14. Transport Information

15. Regulatory Information

16. Other Information

One must approach MSDS sheets with a certain level of chemical awareness and intuition. These data sheets attempt to list all potential hazards associated with a substance, and in so doing can lead to the sense that NO chemicals are safe. The MSDS sheet for water, for example, calls for the following precautions, many of which would undoubtedly cause you some inconvenience the next time you brush your teeth or take a shower!

1. Adequate ventilation is required to protect personnel from exposure to water vapors.
2. Safety glasses with side shields are considered minimum protection.
3. Protective gloves and clothing are recommended when handling water as a vapor or a solid.
4. Water should be protected from freezing, temperature extremes and direct sunlight. In general, water should be stored in a cool, dry, well-ventilated storage room.
5. Emergency eyewash fountains and safety showers should be available in the vicinity of any potential exposure.

This does *not* mean that the information contained within an MSDS is not to be heeded. It simply means that one must use one's scientific and practical training, as well as any published documents, when assessing the potential hazards presented by a compound. In addition, it is wise to consult other sources to learn as much as possible about the hazards of materials with which you work. A number of these sources are listed in the following section.

4.2   Sources of Information about Chemical Hazards

Many other sources of hazard data can, and should, be consulted. These sources include both published and on-line materials, and it is through appropriate consultation with these sources that one can hope to achieve a realistic picture of chemical hazards. The following provide a good starting point for such investigations.

Information about Specific Compounds

- *Chemical and Other Safety Information*, The Physical and Theoretical Chemistry Laboratory Oxford University (http://physchem.ox.ac.uk/MSDS/) – This site provides comprehensive listings of many hazardous chemicals and access to over 15,000 Materials Safety Data Sheets (MSDS).

- *Hazardous Chemicals Database*, The Department of Chemistry, University of Akron (http://ull.chemistry.uakron.edu/erd/) – A searchable database providing detailed listings of health, fire, and other hazards of chemicals, as well as useful chemical and physical properties, including melting and boiling points, vapor pressure, water solubility, and refractive index.

- *The Hazardous Substances Data Bank* (http://toxnet.nlm.nih.gov/) – Maintained by the National Library of Medicine, this Web site provides a wealth of information on health, environmental, and occupational hazards, including numerous references to the primary literature. Comprehensive coverage of available animal testing data, pharmacology, environmental fate, and safety and handling procedures is also provided.

- *International Chemical Safety Cards* (http://www.cdc.gov/niosh/ipcs/icstart.html) – Generated by the World Health Organization/International Labor Organization/United Nations Environment Program and provided on the Web pages of the International Occupational Safety and Health Information Center, these cards "summarize essential health and safety information on chemicals for their use at the 'shop floor' level by workers and employers in factories, agriculture, construction and other work places." This information is meant to complement that contained in the MSDS, providing key information concisely and in a manageable form.

- *Solv-DB®* (http://solvdb.ncms.org/SOLV01.htm), The National Center for Manufacturing Sciences solvent database – A comprehensive listing of health and safety, regulatory, and environmental fate data, as well as physical and chemical properties for a wide range of organic solvents and related compounds.

- *The Merck Index*, 13th Ed., Merck and Co.: Rahway, New Jersey, 2001. The latest edition of an old stand-by, also available in electronic form, the Merck index contains relatively brief but informative descriptions of the properties and hazards of numerous organic compounds.

- *The Sigma-Aldrich Library of Chemical Safety Data*, Lenga, R. E., Ed., Sigma-Aldrich Corp.: Milwaukee, Wisconsin, 1985.

- *Dangerous Properties of Industrial Materials*, 7th Ed., Sax, N. I.; Lewis, R. J., Van Nostrand Reinhold: New York, 1988.

- *Rapid Guide to Hazardous Chemicals in the Work Place*, 2nd Ed.; Sax, N. I.; Lewis, R. J., Eds., Van Nostrand Reinhold: New York, 1990.

- *Fire Protection Guide on Hazardous Materials*, 10th Ed., National Fire Protection Association: Quincy, MA, 1991.

General Laboratory Safety

- *Prudent Practices for Handling Hazardous Chemicals in Laboratories*, National Research Council, National Academy Press: Washington, DC, 1981.

- *Safety in the Chemical Laboratory*, Renfrew, M. M., Ed., Division of Chemical Education, American Chemical Society: Easton, PA, 1967-1991.

- *Safety in Academic Chemistry Laboratories*, 4th Ed., Committee on Chemical Safety, American Chemical Society: Washington, DC, 1985.

Chemical Disposal

- *Hazardous Laboratory Chemicals: Disposal Guide*, Armour, M. A., CRC Press: Boca Raton, FL, 1991.

- *Prudent Practices for Disposal of Chemicals from Laboratories*, National Research Council, National Academy Press: Washington, DC, 1983.

Toxicology

- *Clinical Toxicology of Commercial Products*, 5th Ed., Gosselin, R. E.; Smith, R. P.; Hodge, H. C., Williams and Wilkins: Baltimore, MD, 1984.

- *Carcinogenically Active Chemicals: A Reference Guide*, Lewis, R. J., Van Nostrand Reinhold: New York, 1990.

Environmental Impacts

- *The Hazardous Substances Data Bank* (http://toxnet.nlm.nih.gov/) – Listed under the "Information about Specific Compounds" heading above, this site also provides comprehensive coverage of environmental fate and contamination routes.

- *Solv-DB*® (http://solvdb.ncms.org/SOLV01.htm) – Listed under the "Information about Specific Compounds" heading above, this site also provides information about the environmental fate of a variety of solvents and related compounds.

- *Handbook of Environmental Fate and Exposure Data for Organic Chemicals, Vol. 1: Large Production and Priority Pollutants*, P. H. Howard, Lewis Publishers: USA, 1989.

Summary

Materials Safety Data Sheets (MSDS) provide a wealth of information about the physical and chemical properties of commercially available chemicals, as well as their potential or demonstrated health and environmental impacts. Numerous other sources of information regarding chemical properties and toxicity are also available, both in print and on the Internet.

Chapter 5: Introduction to Green Chemistry

This chapter presents an overview of traditional methods for dealing with and disposing of hazardous chemicals. Following this discussion, the general principles behind green organic chemistry, which seeks to reduce risk by limiting intrinsic chemical hazards rather than merely avoiding exposure, are introduced.

5.1   Traditional Approaches to the Reduction of Environmental Contamination and Human Health Risk

Once a chemical substance has been determined to pose a threat to human health or the environment, what can be done to try to prevent harm? Historically, the strategy has been to try to minimize risk by limiting exposure in the workplace and reducing the quantity of material released into the environment. Established methods to decrease the environmental impact of hazardous waste have focused on reducing emissions through a variety of chemical and engineering means, including:

- use of "scrubbers" to remove volatile compounds from gaseous waste streams
- treatment of waste water to remove organic compounds
- improvement of the efficiency of incineration
- chemical treatment of wastes to reduce toxicity
- minimization of the generation of toxic waste

When applied effectively, these methods can indeed significantly reduce the emission of harmful materials into our environment, and these approaches may be credited with recent successes in reduction of environmental contamination. For example, annual emissions of "dioxins" (TCDD and related compounds) decreased by an estimated 77% between 1987 and 1995. However, as discussed earlier, environmental controls – i.e., controls over the release of hazardous substances rather than over their generation – are subject to failure, as the following two brief but dramatic examples illustrate.

The accidental release of methyl isocyanate in Bhopal, India, in 1984 killed 3,800 people and permanently disabled another 2,700.

The Chernobyl nuclear power plant accident in 1986 led to the evacuation of over 200,000 people and serious contamination of over 10,000 square miles.

Environmental controls are also very expensive – in 1992, American industries spent an estimated $115,000,000 on waste treatment, control and disposal, and in 1996, E. I. DuPont Du Nemours spent $1,000,000,000 (!) on environmental compliance – an amount equaling Dupont's entire budget for research to identify new products!

If, instead of developing methods for controlling the release of hazardous substances, we could *replace these substances with non-hazardous (or, at least, less hazardous) substances*, the problems associated with uncontrolled or accidental release and waste disposal would be greatly diminished or eliminated. Such an approach, in contrast to traditional environmental control approaches, would be both "fail-safe" and of significant economic benefit. *This is the central strategy of green chemistry*, as we discuss in the following sections.

5.2   Strategies of Green Chemistry

How can we make important organic materials with minimal harm to the environment or to our health? Can we use organic chemistry to protect or clean up the environment? The preceding discussion illustrates how chemists and engineers have used their creativity to invent methods to reduce emissions at the *end* of the smokestack or waste pipe and ways to handle hazardous waste to reduce the likelihood of environmental contamination. Green chemistry asks the question, "Can chemists find creative methods that **prevent** the requirement for or generation of hazardous materials in the first place?" In other words, can chemists find new ways to continue the progress we as a society have become dependent upon without causing further harm to the earth and its inhabitants? The answer, as we shall see, is clearly "yes!"

As discussed in Chapter 1, risk is a function of both exposure and intrinsic hazard. Thus, even a comparatively harmless compound can cause health problems in cases of extreme exposure, while a highly toxic substance presents no risk if there is no exposure. It is convenient to formalize this notion in the following equation, representing this functional relationship.

$$Risk = f(\text{exposure, hazard})$$

This equation efficiently makes the point that risk is very small if either exposure or intrinsic hazard is minimized. Green chemistry minimizes risk (both personal and to the environment) by striving to eliminate or reduce the use and generation of hazardous substances, a strategy which contrasts sharply with the traditional approach of focusing on minimizing risk by avoiding exposure. Thus, green chemistry strives to lower the risk, regardless of any potential exposure.

In the most general and global sense, *green chemistry involves inventing new methods to reduce chemical hazards while producing superior products in a more efficient and more economical fashion.* Anastas and Warner [11] have proposed twelve principles to guide the green chemist (Appendix A provides a complete listing of these principles.) Some highlights include:

1. Preventing the formation of waste in the first place
2. Employing safer reagents or solvents
3. Implementing selective and efficient transformations
4. Avoiding unnecessary transformations

Of course, any new chemical transformation or process must still be effective in addition to meeting these criteria. Thus, green chemistry must be the best chemistry – practical and economical as well as green.

---

11. P. T. Anastas and J. C. Warner, "Green Chemistry: Theory and Practice;" Oxford University Press: Oxford, 1998.

In order to obtain a more intuitive grasp of the principles of green chemistry, let's develop a strategy to assess and modify existing reactions with the goal of finding greener alternatives. Any transformation of matter can be written generically according to the following equation.

$$\text{Starting Material} \xrightarrow[\substack{\text{Solvent} \\ \text{Energy}}]{\text{Reagents}} \substack{\text{Products} \\ + \\ \text{Byproducts}}$$

By generalizing in this way, it becomes immediately obvious how we might attempt to make the transformation greener. We can consider reducing the hazards associated with the starting material ("feedstock"), the reagents (substances that act upon the starting material to transform it into the product), and the solvent (a substance, usually liquid, that dissolves the starting material and reagents so that they can react with one another). In addition, we can examine the efficiency of the reaction, considering the energy input (either for heating or cooling the reaction) and the inadvertent production of undesirable substances – the byproducts. Finally, we can evaluate the product itself.

A general strategy for evaluation of a chemical reaction or process and development of greener alternatives is presented graphically below.

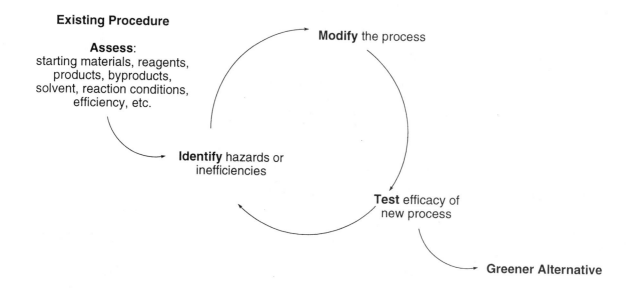

We first analyze the existing process, identifying the materials that are used in the transformation (the starting materials, reagents, and solvents) and those that are produced by the transformation (the

52

product and any byproducts), as well as the reaction conditions (e.g., temperature). We then identify any potential hazards presented by these materials and consider the nature of any energy inputs. Finally, we examine the overall efficiency of the particular reaction or process in question and, more globally, of the whole sequence of reactions or processes being used to generate the final desired product. Consideration of these issues leads us to propose a modified process or procedure, which is then tested to determine whether the new method is effective and whether it reduces hazards. This evaluation and testing will frequently lead to further modification and testing before a final greener alternative procedure is established. By making these considerations an essential part of the design and execution of any chemical process or reaction, they will become part of a chemist's intuition and we will be well on our way toward a greener chemical world.

5.3   Tools of Green Chemistry

In the following chapters, we discuss each of the options available for the design of green chemical reactions and processes. Because solvents and reagents present some of the most significant and large-scale health risks and environmental impacts, we first focus on the choice of alternatives to common hazardous solvents (Chapter 6) and reagents (Chapter 7). Following these discussions, we consider more global issues of reaction design and efficiency, exploring the analysis of reaction efficiency and the development of alternative, greener reactions and processes that minimize or eliminate hazardous waste formation (Chapter 8). We next explore the use of alternative starting materials (feedstocks) and the redesign of products in order to ameliorate health or environmental hazards while retaining (or enhancing) desired chemical or physical properties (Chapter 9).

Summary

Traditional methods for using or disposing of hazardous chemicals have focused on "environmental" controls, attempting to limit exposure to and control the release of chemicals. These methods are prone to failure and accidental release and can lead to unforeseen longer term problems (e.g., the production of TCDD through incomplete combustion of chemical wastes). Recognizing that risk is a function of both exposure and intrinsic hazards of chemical substances, green chemistry seeks to replace hazardous materials with safer alternatives. Each element of a chemical process – starting materials, reagents, solvents, reaction conditions, products, and byproducts – represents a potential target for replacement by a greener alternative.

Chapter 6: Alternative Solvents

This chapter begins with an overview of the most common classes of solvents for chemical reactions, processes, and products, examining the reasons for their popularity and the health, safety, and environmental issues associated with their use. Greener alternatives to the use of these solvents – substitution by safer solvents, modification of processes to allow use of new or unusual solvents, and elimination of the need for solvent altogether – are then considered.

Organic solvents are seemingly ubiquitous in modern industrialized societies. They are essential components of numerous consumer and industrial products, including coatings (paints, inks, etc.), cosmetics (nail polish, hair spray, etc.), household cleaners, automotive and industrial degreasers, and fuels. In addition, they play essential roles in chemical processes, serving to dissolve reactants, assisting in the separation and purification of products, and even facilitating the cleaning of laboratory glassware once a reaction is completed.

As we discuss below, many of the most commonly used organic solvents represent significant health and environmental hazards. Given our heavy reliance on organic solvents and the sheer volume with which they are used, they represent an important opportunity for green chemical thinking.

6.1   Typical Solvents and Problems Associated with Their Use

A remarkable range of organic compounds is used routinely as solvents for various reactions, processes, and preparations. Many of these solvents present significant hazards to human health and to the environment, particularly because most are quite volatile, making exposure and accidental release likely. In the following sections, we briefly review the most commonly encountered classes of organic solvents and their health hazards and environmental impacts.

A. Hydrocarbons

Simple alkanes such as pentane and hexane, as well as more complex hydrocarbons such as those found in kerosene (see Table 6.1) [12] and turpentine [13], are effective solvents for nonpolar compounds, including greases and oils. Liquid fuels, including gasoline and liquefied petroleum gas (LPG), are also comprised of complex mixtures of hydrocarbons.

Table 6.1. Typical kerosene composition [14]

compound class	carbon #	compound	weight percent
Alkyl monoaromatics	10	1,2,3,4-Tetramethylbenzene	1.1
Branched alkanes	10	Isodecane	1.3
	11	Isoundecane	1.2
	12	Isododecane	1.2
	13	Isotridecane	0.9
	14	Isotetradecane	0.6
Monoaromatics	10	Tetralin	0.27
	11	1-Methyltetralin	0.65
	11	2-Methyltetralin	0.68
n-Alkanes	8	n-Octane	3.1
	9	n-Nonane	5.6
	10	n-Decane	5.6
	11	n-Undecane	5.6
	12	n-Dodecane	5.5
	13	n-Tridecane	2.5
Naphthalenes	10	Naphthalene	0.46
	11	1-Methylnaphthalene	0.84
	11	2-Methylnaphthalene	1.8

These hydrocarbons are generally both volatile and flammable and thus represent significant fire risks. In addition, their volatility leads to ready inhalation, either accidentally or intentionally (through

12.   W.E. Coleman, *et al.*, "The Identification and Measurement of Components in Gasoline, Kerosene, and No. 2. Fuel Oil that Partition into the Aqueous Phase After Mixing," *Arch. Environ. Contam. Toxicol.* **1984**, *13*, 171–178.

13.   The major component (65-94%) of turpentine is α-pinene; other common constituents include β-pinene, camphene, myrcene, limonene and β-phellandrene.

14.   ASEAN Review of Biodiversity and Environmental Conservation, November-December, 1999 – on-line journal at http://www.arbec.com.my/.

inhalant abuse), and while there are certainly substances displaying more dramatic health hazards, chronic or acute exposure can lead to significant health problems. Roughly 60,000 cases of hydrocarbon exposure, most involving accidental exposure of young children, leading to perhaps 20 deaths per year, are reported to U. S. poison control centers annually. Central nervous system effects, including disinhibition and narcotic-like depression, are well known and apparently serve as motivation for the "recreational" abuse of hydrocarbons. Lung damage (e.g., pulmonitis, bronchial spasm) is the most common result of hydrocarbon aspiration or inhalation, but sudden death has been ascribed to cardiac dysfunction following extreme hydrocarbon inhalation. Chronic hydrocarbon inhalant abuse leads to a variety of health problems, including cardiomyopathy, cerebellar atrophy, dementia, cognitive deficits and peripheral neuropathy [15].

The volatility of the hydrocarbons represents perhaps their most significant problem with respect to environmental contamination. In addition, hydrocarbons are typically derived from nonrenewable fossil fuel sources and the environmental hazards associated with the extraction of petroleum products (e.g., oil spills, habitat destruction) are well established. Hydrocarbons are released into the atmosphere through evaporation, as well as through incomplete combustion of fossil fuels, and since they are not water soluble, they are not washed out of the atmosphere by rain. As discussed in Chapter 3, sunlight-driven reactions of atmospheric hydrocarbons with nitric oxide (NO) generate ozone (O_3) and nitrogen dioxide (NO_2), contributing to urban haze and causing significant human health and environmental problems.

B. Halogenated hydrocarbons

Halogenated hydrocarbons represent one of the most versatile and powerful classes of solvents. With the exception of chloromethane (methyl chloride), which is not commonly used as a solvent since it is a gas at room temperature (boiling point –24 °C), each of the chlorinated methanes, dichloromethane (methylene chloride), trichloromethane (chloroform), and tetrachloromethane (carbon tetrachloride) is a highly effective solvent for a wide variety of substances. Other halogenated hydrocarbons, including 1,1,1-trichloroethane, trichlorethene, and tetrachloroethene, also experience wide and general use – e.g., as industrial degreasers and dry cleaning solvents – due to their seemingly unique ability to

15.   R. J. Goldstein, "Toxicity, Hydrocarbons," http://www.emedicine.com/ped/topic2721.htm.

dissolve myriad organic compounds as well as their fast evaporation, lack of flash points (the lowest temperature at which a liquid's vapor pressure is sufficiently high to form an ignitable mixture with air near the surface of the liquid) and absence of residue upon drying.

Table 6.2. Volatility of Common Chlorinated Hydrocarbon Solvents

Compound	Boiling Point (°C)	Vapor Pressure @ 20 °C	
		kPa	(mm Hg)
CH_3Cl	-24	489	3,670
CH_2Cl_2	40	46.5	349
$CHCl_3$	61	21.3	160
CCl_4	77	12.0	90.0
$CHCl=CCl_2$	87	7.9	59
CH_3-CCl_3	74	13.4	100
$CCl_2=CCl_2$	122	1.72	12.9

While of exceptional intrinsic value as solvents, the halogenated hydrocarbons in general present appreciable health hazards. As for the simple hydrocarbons, chlorinated solvents are often quite volatile (see Table 6.2), making exposure to significant concentrations of vapors likely. (In order to calibrate yourself with regards to the significance of the reported vapor pressures, recall that water has a vapor pressure of 2.34 kPa, or 17.5 mm Hg, at 20 °C. Thus, nearly all the common chlorinated hydrocarbons are significantly more volatile than water.) Exposure to the chlorinated compounds presented in Table 6.2, as well as many other halogenated compounds not listed, has been found to give rise to a variety of health problems. As highly effective solvents for both polar and nonpolar compounds, the chlorinated hydrocarbons act as "defatting" agents when contacting the skin, leading to irritation and dermatitis. Acute or chronic exposure may also give rise to central nervous system ("CNS"), kidney, liver, and heart problems, as well as immune system suppression. Aspiration, as for the simple hydrocarbons, can cause chemical pneumonitis. Finally, many of the chlorinated hydrocarbons are also suspected of being carcinogenic, mutagenic, and/or teratogenic in humans. In some cases, technical grades of these solvents may contain "stabilizers" that are also carcinogenic.

Hydrocarbons containing both chlorine and fluorine ("chlorofluorocarbons" or CFCs), displaying similar solvent properties to the chlorinated hydrocarbons, represent another common class of halogenated hydrocarbon solvents. Compounds such as trichlorofluoromethane (CFC-11) and dichlorodifluoromethane (CFC-12), more commonly known as "freons™" are highly effective for a variety of applications, including aerosol propellants and refrigeration systems – for example, they are well-known as the active "charges" for automobile air conditioners. Related compounds, used for fire fighting applications, are the "halons", including bromochlorodifluoromethane (Halon 1211, used in hand-held fire extinguishers) and bromotrifluoromethane (Halon 1301, used in "flooding" systems).

Although the freons™ and related compounds present significantly reduced direct human health hazards compared to simple chlorinated solvents, they highlight the essential environmental shortcoming of the halogenated hydrocarbons as a general class of solvents – they contribute to the destruction of stratospheric ozone, as discussed in Chapter 3. Given their volatility, halogenated hydrocarbons and CFCs make their way readily into the upper atmosphere, where solar energy leads to a complex cascade of reactions ultimately resulting in the destruction of the protective ozone layer [16]. Hydrochlorofluorocarbons (HCFCs; i.e., hydrocarbons containing hydrogen as well as chlorine and fluorine substituents) such as chlorodifluoromethane, 1-chloro-1,1-difluoroethane, and 1,1-dichloro-1-fluoroethane appear to be less damaging to stratospheric ozone. However, HCFCs, as well as the CFCs and other halogenated solvents, are potent greenhouse gases, potentially contributing to global warming (see Chapter 3), and are slated in the Montreal Protocol [17] to be phased out by the year 2030.

C. Aromatic hydrocarbons

Aromatic solvents, particularly the "parent" compound, benzene, but also a variety of substituted benzene derivatives, including toluene (methylbenzene), the dimethylbenzenes (xylenes), nitrobenzene, chlorobenzene, and benzonitrile, are commonly used for a wide range of chemical reactions and processes. Many of these compounds are quite volatile (see Table 6.3), and each presents

---

16. F. Sherwood Rowland, Paul Crutzen, and Mario Molina were awarded the Nobel Prize in Chemistry in 1995 for discovering the threat chlorofluorocarbons pose to the ozone layer, for working to preserve the ozone layer, and for protecting human welfare.
17. The *Montreal Protocol on Substances that Deplete the Ozone Layer* is an international agreement originally signed in 1987 and substantially amended in 1990 and 1992. The Montreal Protocol calls for phasing out the production and consumption of compounds that deplete ozone in the stratosphere. I. H. Rowlands, "The fourth meeting of the parties to the Montreal Protocol: Report and reflection," *Environment* **1993**, *35 (6)*, 25-34.

an array of health hazards, as exemplified by benzene itself [18]. Short-term exposure to benzene vapors can lead to typical symptoms of hydrocarbon inhalation, as well as CNS problems, while the defatting action of liquid benzene leads to skin problems akin to those of the halogenated hydrocarbons. Chronic exposure may lead to liver and immune system dysfunction and has been positively linked to aplastic anemia and leukemia. These blood-related diseases may result from benzene-induced cell toxicity in bone marrow cells. In addition, benzene is a known carcinogen and appears to be teratogenic. Given both the widespread use of benzene and the rather dramatic hazards it presents to human health, the biochemical cycling of benzene has received intense scrutiny. While a detailed analysis is well beyond the scope of this discussion, it appears that benzene is oxidatively degraded by the liver, affording a number of metabolites, including muconaldehyde (2,4-hexadiendial) a reactive compound known to link covalently to both DNA and proteins and a recognized cancer causative agent [19]. (Compare this biochemical risk resulting from oxidation of benzene by the liver to the comparable hazards presented by the oxidation of polycyclic aromatic hydrocarbons, as discussed in Chapter 2.)

Table 6.3. Volatility of Common Aromatic Hydrocarbon Solvents

Compound	Boiling Point (°C)	Vapor Pressure @ 20 °C	
		kPa	(mm Hg)
C_6H_6	80	10	75
$C_6H_5CH_3$	111	2.9	22
o-$C_6H_4(CH_3)_2$	144	0.7	5.2
m-$C_6H_4(CH_3)_2$	139	0.8	6.0
p-$C_6H_4(CH_3)_2$	138	0.9	6.8
C_6H_5Cl	132	1.17	8.78
C_6H_5CN	191	0.10	0.75
$C_6H_5NO_2$	211	0.02	0.15

Benzene is routinely released to the environment through a variety of means, foremost of which is the use as a fuel of gasoline, which generally contains roughly 1% benzene. Simple evaporation of

---

18.   See, e.g., The Health Effects Institute, "Program Summary: Research on Benzene and 1,3-Butadiene," March, 1999 (http://www.healtheffects.org/Pubs/benbutps.htm).

19.   L. Latriano, B. D. Goldstein, and G. Witz, "Formation of muconaldehyde, an open-ring metabolite of benzene, in mouse liver microsomes: an additional pathway for toxic metabolites," *Proc. Natl. Acad. Sci. USA* **1986**, *83*, 8356-8360.

gasoline, incomplete combustion of benzene, and conversion of other hydrocarbons to benzene during the combustion process all contribute to its release. Once released, benzene and other aromatic hydrocarbons may be quite rapidly biodegraded by naturally occurring organisms, but if they make their way into anaerobic (oxygen-free) environments (e.g., soil), their degradation is greatly slowed. For example, once *ortho*-xylene leaches into the ground, it undergoes moderate degradation under aerobic conditions (70% degradation after 10 days), but very slow degradation under anaerobic conditions (up to 6 months before onset of degradation). Atmospheric aromatic hydrocarbons react with hydroxyl radicals (HO•) produced by the action of sunlight on various atmospheric components, affording phenols and a number of other compounds presenting potential health and environmental hazards.

D. Alcohols

A variety of alcohols, including methanol, ethanol, *n*-propanol, and *iso*-propanol, are commonly used as solvents and, as a class, are relatively green solvents. Because these solvents are typically volatile (Table 6.4) and flammable, however, they do present some safety issues. Exposure to the vapors of any of the alcohols in high concentrations or over prolonged periods of time can irritate the eyes and lead to CNS effects, including irritation, headache, fatigue, and lack of concentration. Ingestion of methanol can lead to blindness, while chronic ingestion of ethanol is linked to cirrhosis of the liver.

Table 6.4. Volatility of Common Alcohol Solvents

Compound	Boiling Point (°C)	Vapor Pressure @ 20 °C	
		kPa	(mm Hg)
CH_3OH	65	12.3	92.2
CH_3CH_2OH	79	5.8	44
$CH_3CH_2CH_2OH$	97	4.4	33
$(CH_3)_2CHOH$	83	4.4	33

An estimated 86,155 tons of methanol was released in the United States in 1993 (based on the 1993 Toxics Release Inventory). The paper and related products industry accounted for roughly one half of these methanol emissions, while the chemical and allied products industry accounted for about a quarter. In part, this relatively large release volume is related to the comparatively low recognized environmental impacts of methanol and the other simple alcohols.

E. Ethers

Ethers – primarily diethyl ether and tetrahydrofuran – are frequently encountered solvents for organic chemical reactions and processes. They effectively dissolve many classes of organic compounds and are easily removed by evaporation. The high volatility (Table 6.5) of the ethereal solvents, however, makes them appreciable fire or explosion hazards. In addition, the ethers frequently form unstable peroxides on prolonged standing, and this represents a significant explosion hazard. Diethyl ether was once used as an anesthetic, although this usage has been discontinued due to the extreme fire hazard. Recognized health effects of the ethers are not extraordinary – vapors can be irritating to the eyes, skin, and respiratory system, and prolonged exposure can cause dermatitis, narcosis, and, in the case of tetrahydrofuran, liver damage.

$$H_2CH_3C-O-CH_3CH_2 \qquad \qquad \qquad H_3C-\overset{\overset{\displaystyle CH_3}{|}}{\underset{\underset{\displaystyle CH_3}{|}}{C}}-O-CH_3$$

| diethyl ether | tetrahydrofuran | methyl t-butyl ether (MTBE) |

Molecular structures of common ethers

Table 6.5. Volatility of Common Ethers

Compound	Boiling Point (°C)	Vapor Pressure @ 20 °C	
		kPa	mm Hg
$CH_3CH_2OCH_2CH_3$	35	58.6	440
$(CH_3)_2CHOCH(CH_3)_2$	69	15.9	119
$(CH_3)_3COCH_3$	55	32.7 (25 °C)	245 (25 °C)
THF	66	19.3	145

Typical ethereal solvents are not recognized as having severe environmental impacts. However, as new information is obtained, unrecognized impacts do come to light. The case of methyl *tertiary*-butyl ether (MTBE) is a dramatic example. MTBE was developed as a replacement for tetraethyllead as an antiknock agent (octane enhancer) in gasoline, and has more recently been used as an "oxygenate" to facilitate more complete combustion. MTBE is produced in very large quantities (over 200,000 barrels per day in the U.S. in 1999). However, recent studies have detected the widespread presence of MTBE in ground water, including sources of drinking water. In addition to lending an offensive taste, exposure through ingestion may represent a potential human health risk. Animal studies also have shown that inhalation in high concentrations can lead to a variety of health problems [20].

F. Dipolar aprotic solvents

Popular for their high dissolving power are a variety of dipolar aprotic solvents, including the ketones [e.g., 2-propanone (acetone) and 2-butanone (methyl ethyl ketone, MEK)], acetonitrile, N,N-dimethylformamide (DMF), dimethyl sulfoxide (DMSO), and hexamethylphosphoric triamide (HMPA).

Molecular structures of common dipolar aprotic solvents

The ketones are of comparatively low toxicity, presenting health hazards similar to those of the ethers. Their volatility (Table 6.6), as for the ethers, leads to potential fire hazards and results in significant release to the environment through evaporation. Acetonitrile may inhibit cellular respiration, resulting in impaired functions, and exposure at high levels can be fatal. Insidiously, its effects may not appear immediately after acute exposure. DMF can impair the function of the liver and kidneys, and may have

---

20. Studies allowing establishment of any link between MTBE and human cancer are lacking. Limited animal studies have suggested that classification of MTBE as a possible carcinogen may be appropriate, but these studies are still

toxic effects upon human reproduction. In addition, it can enhance the absorption through the skin of compounds dissolved in it, potentially increasing their toxicity. DMSO also enhances skin absorption of solutes. Although it has been touted as a topical treatment for a variety of ailments, chronic exposure can cause skin sensitization, dermatitis, and liver dysfunction. While quite nonvolatile, HMPA presents a host of hazards, including impaired function of the lungs, kidneys, and central nervous system upon short-term exposure. Chronic exposure is linked to kidney and bone marrow effects and to genetic damage and cancer in humans.

Table 6.6. Volatility of Common Dipolar Aprotic Solvents

Compound	Boiling Point (°C)	Vapor Pressure @ 20 °C	
		kPa	(mm Hg)
2-propanone	56	24	180
2-butanone	80	10.5	78.8
acetonitrile	81	9.60	72.0
DMF	153	0.49 (25 °C)	3.7 (25 °C)
DMSO	189	0.0594	0.446
HMPA	232	0.004	0.03

6.2   Traditional Approaches to Solvent Handling and Disposal

In the preceding section, we considered some of the useful properties, health and safety issues and environmental impacts related to various classes of solvents. Given simple structure/activity relationships, we might anticipate that other solvents will present health and environmental impacts similar to those displayed by other members of their solvent class. Thus *alcohols and, to some extent, hydrocarbons and some of the dipolar aprotic solvents, in general represent greener candidates, while chlorinated hydrocarbons and aromatic solvents are generally less green.* In the following section, we will explore, to a greater extent, how to evaluate the greenness of different solvents and, as appropriate, how to find greener replacements for them.

---

under evaluation. U. S. Environmental Protection Agency, "Assessment of Potential Health Risks of Gasoline Oxygenated with Methyl Tertiary Butyl Ether," Office of Research and Development, US EPA, EPA/600/R-93/206.

Various forms of environmental controls are typically used to deal with the hazards presented by organic solvents. Volatile solvents are used in fume hoods to protect the user, and hood effluent is either discharged directly into the atmosphere or treated ("scrubbed") before release. Waste solvents are generally burned (often as fuel in specialized power generation facilities), incinerated, or recycled. Although recycling seems an obvious way to reduce waste, it is often impractical due to technical difficulties or energy costs associated with separating and purifying solvents from spent reaction mixtures.

In our earlier discussions, we have considered environmental control methods and noted that they present a number of potential problems. Accidental exposure through equipment failure or operator error, environmental contamination by untreated fume hood effluent, waste generation from treated discharges, generation of greenhouse gases by burning or incineration, and generation of hazardous substances by incomplete combustion all represent shortcomings of traditional environmental controls.

6.3   Green Approaches to Solvents

Rather than striving to minimize human exposure to and environmental release of hazardous solvents, green chemistry seeks to replace hazardous solvents with alternative solvents or reaction media presenting reduced health risks and environmental impacts. In order to represent a plausible replacement for an existing solvent, an alternative solvent must meet a number of criteria. All green issues aside, the solvent, first and foremost, must successfully dissolve the required starting material and reagents, allowing them to react in the desired manner. Of course the solvent must be inert so that it does not interfere with the desired reaction. Finally, the choice of reaction medium will influence the rate of the reaction, as highlighted in Experiment 10. Once a solvent is chosen which meets these basic selection criteria, it may be further judged by considering the following questions.

1. How efficiently does the reaction proceed in the new solvent? Are undesired side reactions, leading to byproducts, enhanced or diminished by the solvent?
2. Is the solvent nontoxic, or of reduced toxicity with respect to the solvent it is replacing?
3. Does the solvent represent less of a threat to the environment?
4. Does the solvent have lower volatility, leading to reduced emissions or exposure?

5. Is the solvent water-soluble? If so, will this simplify or complicate product isolation and purification, and will it cause difficulties in solvent recycling and/or reuse?

6. Is the solvent soluble in nonpolar solvents? If so, does it accumulate in body fat or tissues?

7. Is the solvent biodegradable? If so, are the biodegradation products safe?

8. Is the solvent available in pure form, or does it contain potentially hazardous impurities?

It is important to realize that many of these questions may be difficult to answer. In particular, our ability to recognize particular compounds or classes of compounds as either safe or of potential risk is limited. As discussed in Chapter 2, ongoing research in the field of "structure/activity relationships" may some day facilitate the discovery or invention of new and safer solvents.

At present, three green approaches to alternative solvent choice are being explored – development of reaction chemistry and processes that allow exploitation of conventional solvents which are known to be relatively safe (e.g., ethanol), conceptualization and design of new, safer solvents, and engineering of new processes that don't require solvents. These approaches are discussed in the following sections.

A. Exploit Greener Conventional Solvents

The combination of intrinsic toxicity and high volatility makes a number of organic solvents particularly hazardous. Three approaches may be envisioned to reduce the health and environmental impacts of solvent use. By avoiding toxic and/or environmentally harmful solvents and relying on relatively nontoxic solvents, we can reduce the intrinsic health hazards associated with solvent use, exposure, and release. By working with less volatile solvents, we can minimize unintentional release or exposure due to evaporation. By recycling solvents, we can avoid the need to dispose of waste solvents, reducing the environmental burden of our chemical activities.

In addition to issues of obvious importance, including *physical properties* such as melting and boiling points, a number of other factors must be considered when choosing a solvent. Table 6.7 provides typical data for a number of common organic solvents, illustrating the types of issues that can be used to guide one's choice of solvent. Consideration of the *volatility* of the solvent, indicated in Table 6.7 by the *evaporation rate* relative to butyl acetate, and the *flash point* (see Section 6.1B) allows one to assess hazards associated with solvent vapor. Determination of *water solubility* can help define any potential for contamination of water supplies upon accidental release, while analysis of the *partition coefficient* for the solvent between *n*-octanol and water ($\log K_{ow}$) can suggest the potential for transport across cell membranes or bioaccumulation in fatty tissues. One measure of the ease of detection of a compound is provided by the *odor threshold* (generally falling in the parts per million range for most organic solvents). The environmental fate of a solvent may be judged in a number of ways, including the *half-life in the atmosphere*, the potential to contribute to urban *ozone formation*, and the time scale of *biological degradation*. Finally, a number of metrics may be used to *quantify* the hazards presented by the solvent. The National Fire Protection Agency (*NFPA*) assigns *ratings* for health, flammability, and reactivity risks on a scale of 0-4, with 0 representing the safest, while the Occupational Safety and Health Administration (OSHA) establishes *permissible exposure limits (PEL)*. LD_{50} values have been determined for many solvents both for ingestion (reported in Table 6.7) and for skin contact, and *specific hazards* such as mutagenicity or carcinogenicity are in some cases well-established. Finally, listing of a solvent in the Environmental Protection Agency's *33/50 program*, which called for phased-in reductions in usage of particularly harmful solvents, can suggest the need for particular concern.

As the data in Table 6.7 indicate, alcohols represent intrinsically rather safe solvents, particularly when compared to other conventional organic solvents. Ethanol and 2-propanol (isopropyl or "rubbing" alcohol) represent particularly good choices. Benzene can be replaced with toluene or xylene, which can cause similar health problems to those of benzene upon acute or chronic inhalation or skin contact, but are appreciably less volatile and appear not to present cancer risks. (For example, toluene is metabolized to innocuous benzoic acid rather than to muconic acid, the highly carcinogenic metabolic oxidation product of benzene.)

Halogenated hydrocarbons represent difficult cases for direct replacement by greener solvents and should be avoided when possible (but see below for further discussions). While a number of replacements have been developed for cleaning and degreasing applications, alternatives applicable for reaction chemistry are sorely lacking. When unavoidable, dichloromethane appears to present the fewest hazards.

Simple ethers such as diethyl ether and tetrahydrofuran may be replaced with polyethers, which display significantly reduced volatility and toxicity while retaining many of the intrinsic desirable solvent properties. For example, 1-methoxy-2-(2-methoxyethoxy)ethane (also known as methoxyethyl ether or "diglyme") has a boiling point of 162 °C and a vapor pressure of only 2.96 mm Hg at 25 °C. Solvents such as ethyl acetate, 1,3-dimethyl-3,4,5,6-tetrahydro-2(1H)-pyrimidone ("dimethyl propyleneurea," DMPU), 1-methyl-2-pyrrolidinone (NMP), and propylene carbonate (PC) represent low-toxicity alternatives to the more hazardous conventional dipolar aprotic solvents.

ethyl acetate   DMPU   NMP   PC

Greener dipolar aprotic solvents

Solvent substitution is, for the most part, immediately achievable and thus represents at the least a sound short-term solution. It is not perfect, of course, as the above discussions of plausible replacement solvents clearly indicates, but it is a start – something that can be done immediately, requiring comparatively little research and development before implementation (i.e., a "drop-in" technology).

Table 6.7: Selected properties of common organic solvents

Solvent	Physical properties		Volatility		Solubility		Detection	Environmental fate			Hazard metrics				
	Mp (°C)	Bp (°C)	Fp (°C)	Evap rate	Water sol (mg/kg)	logK$_{ow}$	Odor thres (ppm)	Atm half life	Urban ozone	Bio deg time	NFPA rating H,F,R	OSHA PEL (ppm)	Specific hazards	LD$_{50}$ (mg/kg)	EPA 33/50 listing?
ALIPHATIC HYDROCARBONS															
Hexane	-95.3	68.7	-26	8.9	Negl	3.90	130	2.9d	0.12	d/w	1,3,0	50	M,T	28,170	N
Ligroin		60-80	-26								1,4,0				
Pentane	-129.7	36.1	-40	>1	Negl	3.39	400	4.1d	0.11	d	1,4,0	600	-	-	N
Petroleum ether		35-60	-49								1,4,0				
CHLORINATED HYDROCARBONS															
Carbon tetrachloride	-23	76.7	-	6.0	793.4	2.83	96	47y	0.00	w	3,0,0	2	C,M,T	2350	Y
Chloroform	-63.5	61.2	-	10.45	7950	1.97	85	160d	0.00	w	2,0,0	2	C,M,T	908	Y
Dichloromethane	-94.9	39.6	-	14.5	13030	1.25	250	110d	0.00	d/w	2,1,0	500	C?,M,T	1600	Y
AROMATIC HYDROCARBONS															
Benzene	5.5	80.1	-11	5.1	1790	2.13	12	13d	0.03	w	2,3,0	1	C,M,T	930	Y
Toluene	-95.0	110.6	4	1.9	526	2.73	2.9	2.7d	0.24	d/w	2,3,0	100	T	636	Y
ETHERS															
1,2-Dimethoxyethane	-57.8	82	-2		∞	-0.21					2,2,0				
Dioxane	11.8	101.3	12	2.42	∞	-0.27	24	1.5d	0.35	d/w	2,3,1	25	M,T	5700	N
Diethyl ether	-116.3	34.4	-45	33	69000	0.89	8.9	1.2d	0.45	d/w	2,4,1	400	M,T	1215	N
Methyl *t*-butyl ether	-109	55.2	-27	8.14	51000	1.24		5.7d	0.07	d/w	2,4,0	40		4000	N
Tetrahydrofuran	-108.4	66.0	-14	4.72	∞	0.46	2.0	1.0d	0.49	d/w	2,3,1	200	T	1650	N
KETONES															
Acetone	-94.7	56.1	-17	5.59	∞	-0.24	13	71d	0.01	d/w	1,3,0	1000		5800	N
Methyl ethyl ketone	-86	79.6	-9	3.8	223000	0.29	5.4	14d	0.04	d/w	1,3,0	200	T	2737	Y
ESTERS															
Ethyl acetate	-83.6	77.1	-4	3.90	80000	0.73	3.9	10d	0.04	d/w	1,3,0	400	M	5620	N
ALCOHOLS															
Diacetone alcohol	-42.8	168.1	58	0.12	∞	-0.098	0.28	12d	0.06	d/w	1,2,0	50	M?	4000	N
Ethanol	-114.2	78.4	13		∞	-0.235	350	4.9d	0.12	d/w	0,3,0	1000	M,T	7060	N
Methanol	-97.7	64.6	11	2.10	∞	-0.77	100	17d	0.08	d/w	1,3,0	200	T	6200	N
2-Propanol	-88	82.2	12	2.5	∞	0.05	28.2	3.1d	0.14	d/w	2,3,0	400	T	5045	N
MISCELLANEOUS															
Acetic acid	16.7	117.9	39	1.34	∞	-0.17	0.48	22d	0.02	d	2,2,1	10	Corr	3310	N
Acetonitrile	-45	81.6	12.8	5	∞		1143	>30d			2,3,0	40	?	2460	N
Carbon disulfide	-110.8	46.3	-30		2000	1.84	0.1				3,4,0	20	T?		N
Dimethylformamide	-61.1	152.8	58	0.17	∞	-1.01					2,2,0	10		3967	N
Dimethyl sulfoxide	18.5	189	95	0.026	∞	-1.35		6.2h	0.23	d/w	1,1,0		M?	14500	N
NMP	-24.4	202	92	0.06	∞	-0.11		0.78d	0.21	d/w	2,2,0		M,T	3914	N
Propylene carbonate	-54.5	241.7	132	0.005	∞		0.017	3.7d	0.08	d/w					
Pyridine	-42.2	208	20		∞	0.7	<1				3,3,0	5			

Fp = Flash point (°C).
Evap rate = Evaporation rate (relative to butyl acetate).
logK$_{ow}$ = log (Octanol-water partition coefficient).
Odor thres = Odor threshold (ppm).
Atm half life = Atmospheric half-life (h = hours, d = days, y = years).
Urban ozone = Urban ozone formation potential (relative to ethylene).
Bio deg time = Biological degradation time (d = days, w = weeks).
NFPA = National Fire Protection Association ratings for health, flammability, and reactivity (0-4 scale, 0 = safest).
OSHA PEL = Occupational Safety and Health Administration permissible exposure limit (ppm).
Specific hazards: M = mutagenic, T = teratogenic, C = carcinogenic, Corr = corrosive.
LD$_{50}$ = Lethal dose for oral ingestion (mg/kg).
EPA 33/50 listing = Listing in Environmental Protection Agency 33/50 program (33% reduction by 1992, 50% by 1995).

B. Design New Solvents or Find Alternative Reaction Media

The invention of new solvents and reaction media represents one of the frontiers of green chemistry. In this section, we introduce a variety of new greener organic solvents and alternative reaction media (including the "fluorous phase," water, supercritical fluids and ionic liquids). Each of these alternatives offers new opportunities for the reduction of the intrinsic hazards of chemical processes.

i. Safer organic solvents

Using both knowledge and intuition, it can prove possible to modify existing solvents to provide for reduced volatility, toxicity, and environmental hazards or to identify new liquids that can be developed as solvents. Although solvents are often used in such large quantities that requiring their synthesis rather than isolation from natural sources (e.g., petroleum) may seem prohibitive, sufficiently positive health and environmental impacts can indeed provide the motivation for change. Examples of such "engineered" solvents are well known, and indeed some are now so well established that they are included in the above discussion of safer conventional solvents. Table 6.8 provides a representative list of such solvents and some of their relevant physical properties. (Again, recall that water has a vapor pressure of 2.34 kPa, or 17.5 mm Hg, at 20 °C.)

Each of these solvents is in use as a reaction solvent and in a variety of other applications. Propylene carbonate is an effective solvent for a variety of epoxy, urethane and other polymers and resins. N-methyl pyrrolidinone (NMP) dissolves many oils when hot, then releases them upon cooling. Lactate esters, including ethyl lactate, are excellent solvents for many polymers and resins. These esters, produced by fermentation, are also biodegradable and recyclable; the ethyl and butyl esters are even approved as food additives. (Ethyl lactate has a fruity odor and contributes a fruity and buttery taste to artificial butter, cheese, rum, strawberry, and other flavorings; butyl lactate also has a fruity odor and a caramel taste with mushroom undertones.) Diacetone alcohol serves as an effective replacement for acetone, dissolving a wide variety of fats, oils, waxes, and resins while retaining water miscibility. While appreciably less volatile than acetone, diacetone alcohol may be recycled by distillation.

Wellcraft Marine, in Sarasota, Florida, has provided a dramatic example of the concept of solvent modification. Even with recycling, Wellcraft was using large quantities of acetone for fiberglass resin cleanup, losing the majority of the acetone through evaporation. By replacing acetone with diacetone alcohol, solvent emission through evaporation was virtually eliminated, and the cost savings quickly covered the expense of a required new recycling apparatus. In order to take a second step toward green processing, Wellcraft is exploring replacement of diacetone alcohol with propylene carbonate, a still safer solvent in part due to its lower volatility [21].

Table 6.8. Representative solvents designed for reduced toxicity and environmental impact

Name	Structure	mp (°C)	bp (°C)	Vapor Pressure @ 20 °C	
				kPa	mm Hg
Propylene carbonate		-55	242	0.0004	0.03
NMP (N-methyl-2-pyrrolidinone)		-24	202	0.039	0.29
DMPU (1,3-dimethyl-3,4,5,6-tetrahydro-2(1H)-pyrimidinone)		-20	146 (@44 mm Hg)	low	low
Ethyl lactate (ethyl 2-hydroxypropanoate)		-25	154	0.23	1.7
Diacetone alcohol (4-hydroxy-4-methyl-2-pentanone)		-44	172	0.13	0.95

ii. Fluorous phase

The halogenated hydrocarbons, discussed above, offer many chemical advantages as solvents, and the chlorofluorocarbons (CFCs) are valuable agents for refrigeration and air conditioning. In response to the environmental impacts of the CFCs, intense research led to the development of hydrofluorocarbons (HFCs). These compounds retain the essential properties required for a CFC replacement and, because they contain no chlorine, do not present risk of damage to the ozone layer in the upper atmosphere.

---

21. "Envirosense - Fact Sheet: New Cleaning Solvents for Industrial Cleaning," http://es.epa.gov/techinfo/facts/florida/newslvfs.html.

Finally, the HFCs are quite safe, and the high solubility of oxygen in HFCs combined with their low toxicity has even led to the exploration of HFCs as "synthetic blood" and as breathable liquids [22].

While the hydrofluorocarbons (HFCs) offer promise as nontoxic, environmentally benign alternatives to chlorofluorocarbons (CFCs), they are not yet of general utility as solvents for organic reactions. This is largely because the solvent properties of HFCs differ dramatically from those of their hydrocarbon analogues, as is vividly illustrated by the immiscibility (mutual insolubility) of tetrafluoromethane and methane [23]. Chemists are currently exploring ways to exploit these unusual properties to permit the use of HFCs as reaction solvents and as media for the separation and purification of reaction products.

Professor Dennis Curran, at the University of Pittsburgh, has explored the synthesis of fluorinated analogues of conventional organic solvents and reagents, including $(C_6F_{13}CH_2CH_2)_3SnH$. By maintaining a CH_2CH_2 linkage between the tin center and the fluorinated alkyl chain, the reagent retains the intrinsic reactivity of simpler reagents like tributyltin hydride $[(CH_3CH_2CH_2CH_2)_3SnH]$. In contrast to reactions using the latter, which frequently demand tedious work-up to remove tin byproducts, separation of the spent fluorinated tin reagent from the reaction mixture is trivial, being accomplished with a simple extraction with a fluorinated hydrocarbon solvent. In a similar vein, Curran has prepared materials for use in column chromatography that are surface functionalized with fluorinated side chains. Separation of mixtures of organic products and fluorinated byproducts on such chromatography columns is so trivial as to beg the definition of chromatography, acting phenomenologically more like a simple filtration. Thus, elution with a conventional organic solvent removes the organic compounds, and then elution with a fluorinated solvent brings the fluorous components down.

iii. Water

Water is often viewed as the single best solvent in terms of minimization of health hazards and environmental impacts. However, water is an infrequently used solvent for organic chemistry, for the simple reason that the vast majority of organic compounds are not appreciably soluble in water. In

---

22. For a recent discussion, see: H.-J. Lehmler, P. M. Bummer. And M. Jay, "Liquid ventilation - A new way to deliver drugs to diseased lungs?" *Chemtech* **1999**, *29(10)*, 7-12.

addition, many organic reagents and intermediates (e.g., $SOCl_2$, Grignard reagents) are reactive toward water, requiring use of anhydrous (water-free) organic solvents. Over the past several years, enormous strides have been made in addressing both of these problems [24]. As but one example, Grignard-type addition reactions can now be carried out in aqueous solution (or suspension, in the case of water-insoluble reactants), using indium metal in place of magnesium [25].

An example of a Grignard-type reaction carried out in aqueous solution

Particularly noteworthy advances have been made in the area of metal-catalyzed reactions. Analogous to the case of fluorinated solvents, advances in this area have required the development of catalysts that display appropriate solubility in water. For example, the catalytic hydrogenation of an alkene to the corresponding alkane, commonly effected in organic solvents by metal-based catalysts like "Wilkinson's catalyst" – $[(C_6H_5)_3P]_3RhCl$, in which each phosphorus atom donates a pair of electrons to the rhodium atom – can be carried out in water with an analogue of Wilkinson's catalyst bearing pendant groups designed to enhance its water solubility [26]. You will have the opportunity to explore the use of a water-soluble catalyst in Experiment 13.

A water-soluble catalyst for catalytic hydrogenation

23. R. L. Scott, "The Anomalous Behavior of Fluorocarbon Solutions," J. Phys. Chem. 1958, 62, 136-145; J. H. Hildebrand, J. M. Prausnitz, and R. L. Scott, "Regular and Related Solutions;" Van Nostrand Reinhold Company: New York, 1970.

24. See, e.g.: C. J. Li and T. H. Chan, "Organic Reactions in Aqueous Media," Wiley Interscience: New York, 1997.

25. C. J. Li and T. H. Chan, *Tetrahedron Lett.* **1991**, *32*, 7017.

26. J. M. Grosselin, C. Mercier, G. Allmang, and F. Grass, *Organometallics* **1991**, *10*, 2126-2133.

An intriguing outcome of the use of water as a solvent is that, in some cases, it can accelerate chemical reactions. Often, this acceleration appears to be due to the water insolubility of the reactants – by virtue of their hydrophobicity, the reactants are forced together, facilitating formation of the transition state for the reaction. A dramatic example is provided by the Diels-Alder reaction of butenone with cyclopentadiene, which is some 700 times faster in water than in isooctane (2-methylheptane) [27].

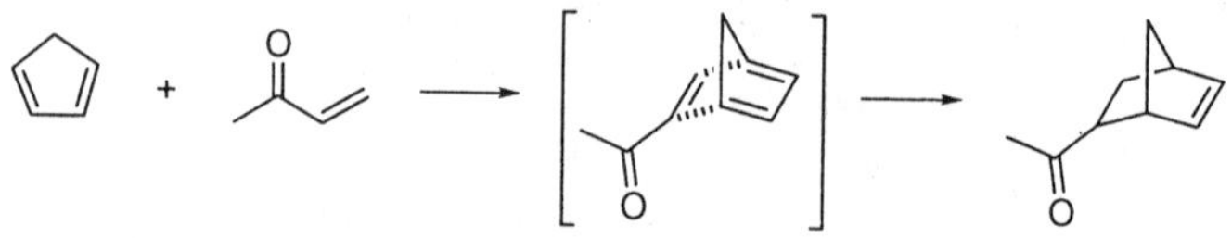

The Diels-Alder reaction between cyclopentadiene and butenone

It is important to note that, whereas water is indeed a remarkably safe solvent, it can be more difficult to purify than many organic solvents. Furthermore, any impurities or contaminants released in aqueous waste streams will, by their very nature, readily find their way into aquifers, enhancing the risk of human exposure. Thus, although water displays many attractive features as a solvent, it is not without its own set of problems. This provides a simple but significant example of the potential complexity in green decision-making, a matter discussed in detail in Chapter 10.

iv. Supercritical solvents

We are all aware that carbon dioxide is a gas under standard conditions of temperature and pressure, and that solid carbon dioxide ("dry ice") sublimes – i.e., it evaporates without melting. Under appropriate conditions of temperature and pressure, however, carbon dioxide can be maintained in the liquid state, as illustrated by the phase diagram for carbon dioxide (shown below). This phase diagram reveals a fundamental property of all liquids. Note that the line separating liquid CO_2 from gaseous CO_2 terminates at a certain pressure and temperature. At this point, called the critical point, there are no longer two phases, but rather a single phase that displays properties intermediate to those of a gas and a liquid. If the temperature and/or pressure are raised above the critical point, a fluid is by definition called a supercritical fluid.

---

27.   D. C. Rideout and R. Breslow, *J. Am. Chem. Soc.* **1980**, *102*, 7816.

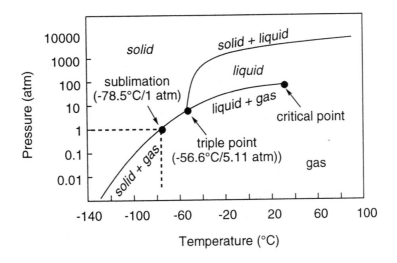

Pressure-Temperature phase diagram for carbon dioxide

Supercritical carbon dioxide is a most unusual solvent, dissolving compounds like a conventional solvent yet penetrating into tiny pores and cracks like a gas. It holds great promise as a replacement solvent for many applications, including dry-cleaning, cleaning of electronic and industrial parts, decaffeination of coffee (50 million pounds/year!), nicotine removal from tobacco, essential oil isolation, and chemical reactions. The critical point of carbon dioxide is at a pressure of 73 atmospheres (7,400 kPa) and a temperature of 31.1 °C. Typical applications using supercritical CO_2 operate at temperatures between 32 °C and 49 °C and pressures between 73 atmospheres (7,400 kPa) and 238 atmospheres (24,100 kPa). Product isolation from supercritical CO_2 is particularly simple – simply releasing the pressure allows the CO_2 to convert to the gaseous state (which may be recovered and reused) and releases the dissolved compound, with no rinsing or drying required. As is the case for the fluorous phase, research has been and remains required in order to exploit the unique solvent properties of supercritical CO_2. As but one example, chemists are studying the synthesis of surfactants (essentially soaps) to solubilize organic substances in supercritical CO_2.

Although the greatest attention has been placed on supercritical CO_2, other supercritical solvents also display unusual properties that bear further exploration. For example, water has a critical point of 218 atmospheres (22,100 kPa) and 374 °C. Interestingly, these conditions may be met geologically, and supercritical water may play a role in mineral deposition and other geological processes.

v. Ionic liquids

Salts like sodium chloride and ammonium nitrate are crystalline solids at room temperature. Interestingly, through the years, various mixtures of salts have been discovered that melt at or near room temperature, and such "molten salts" represent unusual reaction media. Recently, research attention has been focused on organic salts that are liquids at or near room temperature.

N-dodecylpyridinium chloride
mp 68 °C

1-ethyl-3-methylimidazolium acetate
mp -45 °C

Typical ionic liquids

Such "ionic liquids" display a number of unusual properties that suggest they may be attractive alternative solvents for chemical reactions and separations [28]. Most notably, they are essentially nonvolatile, eliminating risks of human exposure or environmental contamination due to evaporation. Their useable liquid range (i.e., the range of temperatures between their melting points and their boiling or decomposition points) is typically around 300 °C, providing much greater versatility in terms of reaction temperature than many common solvents (e.g., toluene – useable liquid range of 204 °C, ethanol – 208 °C, water – 100 °C). Importantly, ionic liquids are remarkably effective at dissolving a range of organic, inorganic, and polymeric materials. They are in general quite easy to prepare and, as a result, they are relatively cheap. Chemists worldwide are actively exploring the translation of organic chemical reactions and processes, including stoichiometric and catalytic reactions, electrochemical transformations, and product isolation, from traditional solvents to ionic liquids. As but one example, Professor Kenneth Seddon, at the Queen's University, Belfast, Northern Ireland, has found that conventional extraction solvents may be effectively replaced by ionic liquids for the recovery of an antibiotic, erythromycin A, produced in laboratory fermentation cultures [29].

---

28.   See, e.g.: http://www.ch.qub.ac.uk/resources/ionic/review/review.html.

76

erythromycin A

Molecular structure of the antibiotic erythromycin A

Despite considerable recent enthusiasm for ionic liquids as greener reaction media, a number of important questions remain unanswered. Little is known about their toxicity, they can under some conditions decompose into volatile toxic materials, and the separation of products from these reaction media may require the use of more conventional volatile organic solvents.

C. Engineer New Processes Not Requiring Solvents

Although great progress has been made in the development of alternative solvents, one should not lose sight of an even better alternative – the elimination of solvents altogether! Solvents are typically used to allow the intimate mixing of chemical reactants and to facilitate the separation and purification of reaction products. Particularly in larger scale processes, solvents also provide an efficient vehicle for the removal of energy from exothermic reactions, whether through direct heat transfer to the surrounding medium or by boiling, transferring heat to a cooling tower. While certain applications will most likely always require a solvent, when analyzing any particular reaction or process for greenness, it is well worth considering whether one can do away with the solvent entirely. Some reactions can be effected in the solid state (as illustrated by Experiments 15 and 17), while others can be carried out with vaporized (gas phase) reactants (as in Experiment 7). As is the case for supercritical fluid and ionic liquid technologies, more research is needed to determine the generality of solid-state and gas-phase reactions.

---

29. S. G. Cull, J. D. Holbrey, V. Vargas-Mora, K. R. Seddon, and G. J. Lye, "Room-Temperature Ionic Liquids as Replacements for Organic Solvents in Multiphase Bioprocess Operations," *Biotechnology and Bioengineering*, **2000**, *69(2)*, 227-233.

Summary

Organic solvents are used in a spectacularly diverse array of applications, from chemical reactions and processes to personal health care products. Many of the most commonly used solvents, including hydrocarbons, halogenated hydrocarbons (including chlorofluorocarbons), aromatics, ethers, and dipolar aprotic solvents, can display adverse effects on human health and/or environmental quality. In contrast to the traditional use of environmental controls to address these issues, green chemistry seeks to replace these solvents with alternative solvents that are safer yet retain the requisite solvent properties. Current green approaches to solvent selection may be grouped into three categories: 1) substitution of safer conventional solvents (e.g., alcohols) for more hazardous solvents; 2) modification of solvent and reactant structure to allow the use of new solvents, including safer organic solvents (e.g., ethyl lactate, fluorinated solvents, and ionic liquids), supercritical solvents (e.g., carbon dioxide), or water; and 3) elimination of solvents entirely through the engineering of processes that may be effected without them.

Chapter 7: Alternative Reagents

This chapter surveys green approaches to the selection of chemical reagents. The advantages to be gained through the exploration of both milder and untraditional reagents are discussed. Following a survey of modern approaches to reagent selection, the "three R's" of green chemistry – recoverability, recycling, and regeneration – are presented.

7.1   Alternative Reagents

We have defined reagents as those chemical reactants that are used to effect the conversion of a starting material to a product. Many of the reagents that one encounters in a typical organic chemistry textbook, while effective for the chemical transformation being presented, represent health and environmental nightmares. Phosgene, bromine, dimethyl sulfate, hydrogen cyanide, heavy metal salts (e.g., chromium, mercury), ... – the list goes on and on. In addition, many reactions call for prolonged reaction times at elevated temperatures, requiring significant energy inputs that, ultimately, trace back to environmentally damaging combustion of fossil fuels or nuclear power. Green chemistry takes two approaches to dealing with the hazards presented by chemical reagents – the use of safer alternative or nontraditional reagents and the development of more selective and more efficient reactions minimizing the need for hazardous or waste-producing reagents. The former approach – use of safer and nontraditional reagents – is discussed here; the latter – reaction efficiency – forms the basis of the discussion in Chapter 8.

7.2   Safer Reagents

Organic reaction chemistry is replete with examples of the use of highly reactive reagents to carry out simple chemical transformations – the chemical equivalent of using a sledgehammer to push in a thumbtack. These highly reactive reagents are frequently toxic and/or environmentally harmful. Traditional approaches to dealing with these hazards are largely focused on environmental control – avoiding contact, inhalation, or ingestion through use of fume hoods and gloves, avoiding contamination of water, burning or incinerating (along with solvents) if possible, and chemically treating to reduce solubility followed by burying in a landfill.

Finding alternative reagents often means simply finding a milder reagent – one that is sufficiently reactive to carry out a desired transformation, but not so reactive that it poses unacceptable hazards. Such an approach is explored in Experiment 5 (the oxidation of cyclohexene to adipic acid). In this experiment, hydrogen peroxide is used in place of nitric acid, a powerful but hazardous, pollution-generating, and nonselective oxidizing agent. Choosing reagents with only the required "horsepower" to effect a desired transformation allows one to obtain greater selectivity while avoiding undesirable reagents. In addition, this approach offers a number of other practical advantages, including the following.

- Formation of byproducts can be minimized, reducing the need for separation and purification steps, each of which represents an additional source of hazards and wastes.

- Requirements for "protecting groups," employed to prevent undesired reactions of functional groups while carrying out desired reactions elsewhere within a molecule, can be reduced.

In order to replace a hazardous reagent with a "greener" reagent, three standards should be considered.

- Efficacy – the alternative reagent must be able to carry out the desired transformation with comparable or superior efficiency to the existing reagent.

- Safety – the alternative reagent should display reduced volatility, flammability, toxicity, and/or reactivity and increased stability relative to the existing reagent. Corrosives, lachrymators, and cancer-causing reagents should be avoided.

- Environmental impacts – the alternative reagent should represent a reduced environmental impact, either upon unintentional release of the reagent itself or upon release of the byproducts or waste from the process utilizing the reagent.

As always, it is important to note that, when approaching green chemistry, even small steps can represent an improvement. Thus, an alternative reagent that does not fulfill each of these standards may still represent a viable option. For example, benzoic acids are often prepared via the oxidation of side chains of alkylbenzene derivatives. This reaction is often effected with an excess of potassium permanganate in strongly basic solution, typically containing an excess of pyridine. The same oxidation may be effected catalytically, using molecular oxygen as the oxidant and a cobalt salt as the catalyst [30]. Although the catalyzed process requires a small amount of HBr and uses acetic acid as a solvent, it avoids the use of $KMnO_4$ and pyridine and generates no heavy metal wastes.

Catalytic side chain oxidation

7.3   Nontraditional Reagents

Thinking "outside the box" can offer opportunities for dramatic conceptual breakthroughs in a variety of settings, including the organic chemistry laboratory. For example, one can imagine a variety of alternative reagents, including electrons, photons, organisms, and enzymes, to effect the myriad reactions essential to organic chemistry. Each of these alternatives is discussed in the following sections, and several of these alternatives are explored in the experimental section of this text.

A. Electrosynthesis.

Reduction and oxidation reactions often utilize reagents that are not very green. Since redox reactions involve the transfer of electrons, "outside the box" thinking might suggest effecting them using a simple electrochemical cell. Thus, one might be able to replace a redox reagent with a pair of electrodes and a battery! Such "electrosynthesis" remains under active research investigation. A dramatic example, illustrating both the efficacy and the potentially unique selectivity of

---

30.   A. S. Hay and N. S. Blanchard, *Can. J. Chem.* **1965**, *43*, 1306; G. J. Powels and B. Zeelie, "Catalysts for the

electrosynthesis, is provided by the electroreduction of pentachloropyridine, which affords symmetrically substituted 2,3,5,6-tetrachloropyridine. This electroreduction process has been used commercially by Dow AgroSciences in the production of "chlorpyrifos" ("Dursban"), a well-known pesticide (e.g., for eradication of termites) [31].

Electrochemical dechlorination of pentachloropyridine, en route to "Dursban"

B. Photochemistry.

Photons (light energy) represent another easy-to-overlook type of "reagent." The energy provided by sunlight, or by more specialized light sources, can be used to make or break chemical bonds. A dramatic example is provided by Experiment 16, in which *trans*-cinnamic acid undergoes dimerization to a cyclobutane derivative upon exposure to sunlight. Not only does this reaction proceed efficiently in the absence of any other reagents, but also it affords access to a compound that would be relatively difficult to prepare in any other way, and for which stepwise synthesis would presumably require a multitude of hazardous reagents. Reactions facilitated by ultrasound waves (sonochemistry) and by microwave energy (see Experiment 8) are also under active exploration.

C. Biological "Reagents"

The use of living organisms or isolated enzymes as "reagents" also represents an ongoing area of research emphasis. If a desired compound can be extracted from a renewable plant resource or generated by fermentation using naturally occurring (or genetically altered) organisms, this natural source may be appreciably greener than any synthetic source.

---

production of aromatic carboxylic acids," RSA Pat. 95/4777, Granted on 24/04/1996.

31. Dursban is no longer approved for usage and is thus a distinctly non-green product. This, however, does not negate this particular transformation as an excellent example of an electrochemically-mediated synthesis.

The promising anticancer agent taxol was originally isolated from the bark of the Pacific yew tree, *Taxus brevifolia*, a very slow growing species. The initial discovery led to rampant destruction of native yew trees as their bark was stripped, leading to quick death of the entire tree. As it quickly became apparent that natural sources of taxol would be woefully inadequate, chemists turned their attention to the synthesis of taxol "from scratch." The complexity of taxol, unfortunately, is so extreme that such "total syntheses" are unlikely to prove profitable. However, a different species of yew tree, the English yew (*Taxus baccata*), was discovered that provides a compound easily converted to taxol. Significantly, this compound (10-deacetylbaccatin III) is found in the needles of the tree rather than the bark, allowing for periodic harvesting without causing the death of the tree [32].

Molecular structures of taxol and 10-deacetylbaccatin III

Penicillin, originally isolated by Alexander Fleming in 1929 from a mold, *Penicillium notatum*, is a well-known antibiotic. An intensive search for other molds producing larger quantities of penicillin followed Fleming's initial discovery. In 1943, Mary Clark found a moldy supermarket cantaloupe, infected with *P. chrysogenum*, and this mold provided twice the penicillin yield of *P. notatum*. As bacteria have developed resistance to penicillin, scientists have sought to prepare analogues. The total synthesis of penicillins, while achievable, is complex and expensive. By carrying out simple chemical transformations of penicillin produced by *Penicillium* molds, however, chemists can easily prepare a wide variety of "semisynthetic" penicillin derivatives. Interestingly, the molds used to produce penicillin are grown in culture broths containing "corn steep liquor," a byproduct and former waste of corn processing, illustrating another key facet of green chemistry – the use of alternative feedstocks, as discussed in chapter 9.

32. K. C. Nicolaou, Z. Yang, J. J. Liu, H. Ueno, P. G. Nantermet, R. K. Guy, E. A. Couladouros, C. F. Claiborne, J. Renaud, K. Paulvannan, and E. J. Sorensen, *Nature* **1994**, 630; J. N. Denis, A. E. Greene, D. Guenard, F. Gueritte-Voegelein, L. Mangatal, and P. Potier, *J. Am. Chem. Soc.* **1988**, *110*, 5917.

Preparation of "semisynthetic" penicillin antibiotics

D. Biocatalysis

The design of efficient and robust catalysts, operating at lower temperatures, offering the opportunity for higher selectivity and reactivity, and adaptable to a variety of reactants, can represent a difficult process and is an extremely active area of research. Ongoing research in the area of biological catalysis is particularly relevant to our studies. By recruiting Nature's catalysts, the enzymes, chemists can often develop highly efficient and selective preparations of desired organic compounds, in aqueous solution and at or near room temperature. Products containing stereocenters can often be obtained in a high state of stereoisomeric purity, further enhancing the utility of biocatalysis. Significant progress has been made in this area, both in the use of purified enzymes to carry out desired transformations and in the use of whole organisms to effect either single step or complex multi-step synthetic procedures.

Professors John Frost and Karen Draths, at Michigan State University, have used genetically altered microbes to create novel routes to medicinal natural products and industrial chemicals utilizing simple carbohydrates such as D-glucose, D-xylose, and L-arabinose as starting materials. The carbohydrates are obtained from plants, representing a completely renewable resource, rather than from petroleum, the nonrenewable source of many traditional organic chemicals. Examples include the microbe-catalyzed synthesis from D-glucose of gallic acid, previously obtained from gall nuts (Couteria tinctoria) and a starting material for trimethoprin, a

widely prescribed antibiotic, and the syntheses of adipic acid, catechol, and vanillin from glucose
instead of benzene.

D-glucose   gallic acid   trimethoprin

adipic acid   catechol   vanillin

Biocatalyzed synthesis of organic chemicals from carbohydrates

E. Catalysis

Replacement of a reagent required in stoichiometric quantities with a reagent that functions catalytically can provide for dramatic improvement in the greenness of a reaction. A catalyst facilitates a chemical reaction without itself being consumed, and thus a catalyzed reaction may require only a very small amount of the catalyst. Green benefits are immediately achievable on at least two fronts. Since a catalyst is used in small quantities and, ideally, is recoverable, a significant reduction in waste production should be realized, and since a catalyst facilitates a reaction, reduced energy inputs (e.g., in heating the reaction) should be possible.

Ideally, a catalyst should greatly accelerate the rate of a reaction, be highly selective for the desired reaction, last forever, and be required in very small quantity. The "turnover number," defined as the number of times the average catalyst molecule carries out the desired reaction before it loses its ability to catalyze the reaction, provides one measure of catalytic efficiency. Thus, a catalyst with a high turnover number can be used in sometimes vanishingly small quantities, while one with a lower turnover number might require use of a significant (though still less than stoichiometric) amount in order to achieve a reasonable reaction rate. Particularly for catalysts based on very expensive metals such as rhodium and platinum, high turnover number is clearly desirable.

Given the intimate relationship between reagent choice and reaction design, further issues regarding alternative reagent selection are addressed in the following chapter. We close this chapter with a brief discussion of more global issues regarding alternative reagent choice.

7.4   The "Three R's" of Green Chemistry

When considering alternative reagents, we should also consider ways in which these reagents can be reclaimed and used again. The "Three R's" of green chemistry – Recovery, Reuse/recycling, and Regeneration – can help guide our considerations.

Recovery refers to the isolation of solvents and spent reagents (including catalysts) following the completion of a reaction or process. Filtration represents a simple and potentially effective recovery method, but the majority of chemical reagents will be soluble in the reaction medium and thus not removable by filtration. Precipitation of a desired compound followed by filtration can be effective, but requires selective and, ideally, quantitative precipitation. Standard separation procedures – distillation, crystallization, and chromatography – are for the most part time-consuming, difficult, and expensive, and in some cases may themselves give rise to additional wastes (e.g., recrystallization solvents, chromatographic supports). As a result, reagents and byproducts have traditionally been recovered only when they are either very valuable or extremely hazardous.

One exception to this generalization is that of solvent recovery. Since many solvents are volatile, they are potentially recoverable by distillation. However, in order to carry out solvent recovery by distillation safely, and to afford recovered solvents of satisfactory purity, careful control of waste streams entering the distillation recovery unit is required. Since the waste will be heated, the possibility of undesired reactions during attempted distillation must be considered – the components of the waste stream must be chemically compatible with one another. The boiling point of the desired solvent must be sufficiently different from that of undesired components, and, often, a distillation apparatus with a high separation efficiency (a high number of "theoretical plates") will be required. Generally, in order to have any chance at effecting a realistic recovery of solvent, waste components must be carefully sorted and segregated, a matter that can be difficult in laboratories in which many different solvents are used, often in the same process.

Effective mechanisms for recovery of reagents and solvents are the key to their **reuse** or **recycling**. Although these terms are often used interchangeably, we will use "reuse" to describe those cases in which a solvent or reagent is useable without purification, and "recycling" for cases in which processing or purification is required before the solvent or reagent can be used. Solvent recycling, generally involving distillation, can be used in some cases, subject to the limitations noted above, but is often not feasible, either practically or economically. Reagent reuse may seem impossible, since reagents are consumed in the course of chemical reactions. However, there is one significant exception to this generalization – a catalyst, by definition, is reused many times in the course of a reaction. In addition, cleverly designed catalysts can be isolated from reaction mixtures and reused or recycled for additional reactions (e.g. Na_2WO_4 in the adipic acid synthesis presented in Experiment 5). This is beneficial both environmentally and economically, given the high cost of many common catalysts, which often contain precious metals such as rhodium or iridium.

Even in those cases in which a reagent is consumed, it may prove possible to effect the **regeneration** of the active reagent. Such regeneration approaches are particularly applicable in cases in which the spent reagent is easily recoverable. One especially attractive approach has been the use of solid-supported reagents – reagents that are either covalently or ionically linked to an insoluble material. The support can be any of a wide variety of materials, either organic or inorganic – e.g., organic polymers, clay, glass beads – as long as it is unreactive under the desired reaction conditions. When designed properly, the reagent attached to such a support is still accessible to reactants in a solution contacting the insoluble support. Once the reaction is complete, a simple filtration easily separates the solution, containing the desired product, from the spent reagent, which is still attached to the solid support. The recovered spent reagent may then be chemically converted back to its active form, and the solid-supported reagent is then ready for reuse. In Experiment 14, the use of a resin-bound oxidizing agent is explored. Many other supported reagents have been developed, including brominating agents [e.g., poly(vinylpyridine) partially functionalized with HBr_3] [33] and reducing agents [e.g., poly(styrene) bearing $N(CH_3)_3^+$ BH_4^- substituents] [34]. Supported reagents capable of functioning in water rather

33.   J. Habermann, S. V. Ley, and R. Smits, *J. Chem. Soc., Perkin Trans. 1* **1999**, 2421.
34.   J. Habermann, S. V. Ley, and J. S. Scott, *J. Chem. Soc., Perkin Trans. 1* **1999**, 1253.

than organic solvents have also been developed. A downside of these supported reagent approaches can be the alteration of reactivity and selectivity relative to the reactions of the unsupported reagent, and this remains an area of active research.

Examples of polymer-supported chemical reagents

Closely related to these approaches is the use of reagents designed to have "tunable" solubility. In many cases, simply changing the reaction solvent by the addition of a second solvent can induce the precipitation of the spent reagent, ready for recovery, regeneration, and recycling. Recently, chemists have been exploring supports that are soluble at some temperatures, but insoluble at others, allowing utilization of simple thermal cycling to recover spent reagent. These approaches have the benefit of maintaining the active reagent in solution (albeit attached to the soluble support), leading to fewer unexpected changes in chemical reactivity resulting from the attached support.

Summary

Many widely used reagents represent overkill with regard to the chemical transformation being effected. Through the exploration of alternative, milder reagents to effect desired chemical transformations, one can avoid the hazards of more reactive reagents, reduce dependence on protecting groups, and effect cleaner reactions, thereby generating reduced quantities of less hazardous wastes. A variety of approaches to effecting chemical reactions are under investigation, including electrochemical synthesis, photochemically driven reactions, biological reagents (both whole organism and purified enzymes), and catalysis by both synthetic and biological substances. Investigations of green alternatives to traditional chemical reagents are guided by the "Three R's" of green chemistry – recovery, reuse/recyling, and regeneration.

Chapter 8: Reaction Design and Efficiency

This chapter presents several measures by which the efficiency of a chemical reaction may be quantified. Guidelines for the development of alternative, greener routes for the preparation of desired compounds are provided.

8.1 Reaction Efficiency

Thus far, we have focused on the design or selection of replacements for more hazardous conventional solvents and reagents. While this approach can clearly lead to the development of safer and environmentally friendlier chemistry, it is by and large focused on making relatively small changes in existing procedures. These small changes often lead to vast improvements over existing procedures. Still, in principle, even larger strides toward green chemistry can be made through the design and engineering of better and more efficient reactions and processes.

One can imagine a "perfect" chemical reaction – one that is completely **selective**, affording only the desired product, highly **efficient**, incorporating all atoms of the starting materials and reagents in the product, and entirely **safe**, using only nonhazardous (to the chemist and to the environment) starting materials, reagents, and products. Ideally, this reaction would require **no solvent** and **no energy inputs** – either heating or cooling from ambient temperature. In this chapter, we will explore the meaning and measure of reaction efficiency and the design of alternative, greener reactions and processes that minimize or eliminate hazardous waste formation.

Chemists generally quantify the efficiency of a reaction by reporting the "chemical yield," which is simply defined as the percent of the theoretically possible number of moles of product obtained. In other words:

Chemical yield = (moles of product obtained/moles of product possible)*100%

As a simple example, consider the following reaction.

	benzyl bromide	sodium ethoxide	benzyl ethyl ether	
molecular weight:	171.038	68.051	136.195	102.894
grams used:	17.1	10.2		
moles used:	0.100	0.150		

The reaction proceeds with the indicated stoichiometry, and thus one mole of benzyl bromide should react with one mole of sodium ethoxide to produce one mole of benzyl ethyl ether. Benzyl bromide is the "limiting reagent," since there are fewer moles of it present than of sodium ethoxide, and therefore the maximum amount of benzyl ethyl ether that could be obtained is 0.100 moles. If one carried out this reaction and obtained 12.67 g (0.093 moles) of benzyl ethyl ether, then the chemical yield of this reaction would be (0.093/0.100)*100, or 93%.

The chemical yield certainly reveals important information about the efficiency of a reaction and is thus an important green consideration. A reaction that proceeds in 99% yield would generally be considered an excellent reaction, while one that affords a product in 2% yield would generally be considered a poor one, since it would be affording 98% of something else (byproducts or, perhaps, recovered starting material). (Note, however, that if the product of this 2% yield reaction could save your life, a 2% yield might look pretty good – certainly better than 0%!)

However, chemical yield is only part of the story. Consider the following reaction, the final step in the synthesis of methylamine via the "Gabriel reaction."

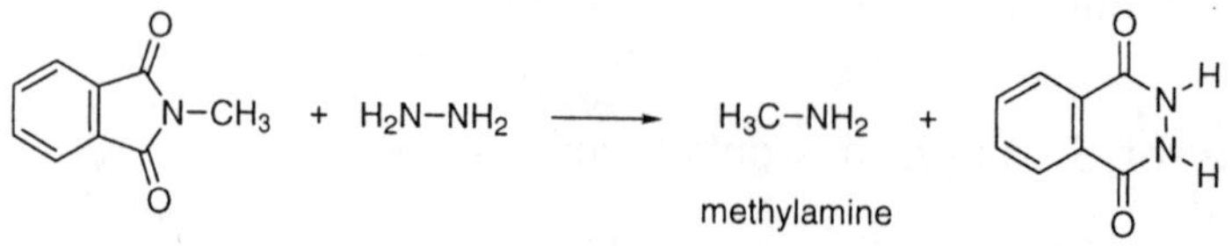

Assume the reaction goes to completion, allowing the isolation of methylamine in 100% yield. Is this an ideal green reaction? No! While the chemical yield is high, the reaction generates an extremely large amount of waste. Of all the atoms going into the reaction ($C_9H_7NO_2 + N_2H_4$), only a small number (CH_5N) are incorporated in the product; the remainder are converted to waste.

The concept of "atom economy" was developed to allow the quantification of the efficiency of utilization of the atoms provided by the starting materials and reagents [35], as opposed to simply the chemical yield. Atom economy is defined as the percent of atomic weight of all the starting materials appearing in the final product. In the above example, the atom economy is (MW_{CH5N})/($MW_{C9H7NO2}$ + MW_{N2H4})*100% = 16.1%. Thus, in this reaction, 83.9% of the molecular mass of the starting materials and reagents is converted to waste! Similarly, the atom economy of the benzyl ethyl ether-forming reaction may be calculated as ($MW_{benzyl\ ethyl\ ether}$)/($MW_{benzyl\ bromide}$ + $MW_{sodium\ ethoxide}$)*100% = 57.0%.

8.2 Experimental Atom Economy

It is important to note that atom economy calculations are based on atom utilization as represented by the stoichiometry of a balanced equation for a chemical reaction. In other words, atom economy only addresses the theoretically possible efficiency of utilization of atoms. Unless one experimentally adheres rigorously to the stoichiometry of the balanced equation, atom economy does not necessarily reflect the *actual* utilization of atoms in a given reaction. In reality, chemists frequently employ one or more reagents in excess, in order to accelerate a reaction or to help drive it to completion.

Let's reconsider the benzyl ethyl ether forming reaction presented earlier, for which we determined a chemical yield of 93% (based on the amount of product obtained) and an atom economy of 57%. While the stoichiometry of the reaction requires only one "equivalent" of sodium ethoxide (i.e., only one mole of sodium ethoxide for each mole of benzyl bromide), we in fact used more than this – 0.150 mole instead of 0.100 mole. As a result, there will be sodium ethoxide left at the end of the reaction, and thus there will be more waste than the atom economy calculation suggests. Professor Michael Cann, of the University of Scranton, has developed the concept of "experimental atom economy", which modifies the atom economy calculation to account for the actual quantities of reagents used

35.   B. M. Trost, "The Atom Economy – A Search for Synthetic Efficiency," *Science* **1991**, *254*, 1471-1477; "On Inventing Reactions for Atom Economy," *Accts. Chem. Research* **2002**, *35*, 695-705.

[36]. The calculation is very similar – one simply compares the theoretically obtainable mass of the desired product to the total mass of reactants utilized (rather than just the amounts called for by the stoichiometry of the balanced equation). In our example, the theoretical yield of benzyl ethyl ether is 0.100 mole, or 13.62 g. The reactants were benzyl bromide (17.1 g) and sodium ethoxide (10.2 g), and thus the experimental atom economy is simply (13.62)/(17.1 + 10.2)*100% = 49.9%.

The ultimate measure of the efficiency of a chemical reaction may be obtained by considering both the chemical yield and the experimental atom economy. This is easily formalized by simply multiplying the chemical yield by the experimental atom economy – in our example, this would be 93% * 49.9% = 46.4%. This has been called the "percentage yield x experimental atom economy." (An alternative measure of the overall efficiency of a reaction is provided by the "environmental quotient," or "E-factor," defined as the total mass of waste divided by the mass of desired product [37].)

Interestingly, if one simply looked at the mass of product obtained and the masses of reactants used and performed a simple yield calculation, forgetting everything one knew about moles, balanced equations, and stoichiometry, one would obtain this same value! This is not to say that these concepts are not important – indeed, they are essential to the planning and execution of chemical reactions – but merely that they are not critical to the analysis of reaction efficiency.

A realistic assessment of the overall efficiency of a reaction, then, must take into account the chemical yield (a measure of selectivity), the atom economy (a measure of intrinsic efficiency), and any deviations from ideal stoichiometry (a measure of the actual efficiency) of the reaction. A good green method must display both selectivity and efficiency, as it is through optimization of both of these measures that waste is minimized and purification is simplified.

8.3   Alternative Reaction Design

Perhaps you have had some experience with editing another student's written work. In the process, you may have found yourself spending far too long trying to reconfigure a written sentence, only to realize,

---

36.   See: http://academic.scranton.edu/faculty/CANNM1/organic.html.
37.   R. A. Sheldon, *Chemtech* **1994**, *(March)*, 38-48.

finally, that it would be much simpler just to discard the sentence and come up with a completely different way of saying the same thing. Reaction design can present the same situation. One can spend days trying to find ways to make a set of reactions or processes greener, sometimes with great success. Alternatively, one can determine what the required result of the set of reactions is – in organic chemistry, this is often the synthesis of a particular compound – and develop an alternative way to achieve it. Though both approaches offer promise, the latter, as in editing, will perhaps more often result in real conceptual breakthroughs. The approaches we have discussed thus far have by and large focused on "rewriting an existing sentence." Let us now briefly consider the alternative – coming up with a completely new "sentence."

8.4   Alternative Synthetic Pathways

Can a desired compound be obtained through an alternative set of transformations that use less hazardous starting materials, reagents, or solvents, that generate less waste, and that consume less energy? This, ultimately, is the "big question." Whether or not one can answer it depends in part on one's creativity, but in large measure on one's knowledge of organic reaction chemistry. Green alternatives may involve the application of known chemical reactions, the modification of existing procedures, and the discovery of new reactions. Guidance in the selection of green alternatives is conveniently provided by the concept of atom economy, which leads to the following immediate and important observations.

Addition reactions are excellent candidates for green reactions. By definition, these are reactions in which the atoms provided by both the starting material and the reagent end up in the product. Thus, addition reactions are intrinsically of very high atom economy. Examples of addition reactions include the Diels-Alder reaction, other cycloaddition reactions, and 1,4 (conjugate) addition to α,β-unsaturated carbonyl compounds.

Diels-Alder cycloaddition   1,4-addition

Catalysis can be an excellent approach to green reaction design. Since catalysts are used in less than stoichiometric amounts, they in general detract little from the experimental atom economy of the reaction being catalyzed.

Substitution reactions can be good candidates for green reactions, depending of course on just what is being replaced by what. For example, electrophilic aromatic nitration results in the loss of only a single hydrogen atom from the substrate. On the other hand, nucleophilic displacement of a tosylate by a bromide ion is a comparatively poor reaction in terms of atom economy.

electrophilic aromatic nitration

nucleophilic displacement of a tosylate

Elimination reactions are poor in terms of atom economy. In essence, they are the reverse of addition reactions, and in contrast to the efficiency of addition reactions, elimination reactions are efficient primarily at producing significant quantities of waste in addition to the desired product.

Protecting groups are terrible in terms of atom economy. The reasons for this poor economy are simple – use of a protecting group requires a chemical reaction to install it, and another to remove it once its job is done. For virtually all protecting groups, the latter, removal of the group, represents a source of considerable waste. Thus, use of a protecting group adds two chemical steps to a process and is guaranteed to generate waste. Although sometimes unavoidable, use of a protecting group is essentially an admission of defeat in efficient synthetic design.

94

Within this context, other considerations also are of significance when determining how green an alternative reaction might be.

- Is the reaction mediated by reagents in stoichiometric amounts, or does it exploit catalysis?

- What is the extent of reaction (i.e., does the reaction go to completion, or is it an equilibrium process that affords, at best, a small amount of the desired product?

- Does the reaction proceed at room temperature, or does it require energy inputs in the form of heating or cooling?

- Is the product obtained easily in pure form, or are complicated and waste-producing separation and purification steps required? A requirement for such separation and purification steps suggests that byproducts are being generated, leading to concerns not only about inefficient reactions and waste generation, but also about potential health hazards or environmental impacts of the byproducts, which may not have been fully investigated.

Summary

The ideal chemical reaction would be efficient, selective, and safe while requiring no solvents or energy inputs. In reality, few if any such ideal chemical reactions are known. Traditionally, the efficiency of a chemical reaction has been measured by the chemical yield, but this calculation ignores both the issue of waste generation and the use of excess reagents to drive reactions to completion. The concepts of atom economy and experimental atom economy have been developed to accommodate these efficiency issues. Development of alternative synthetic pathways to desired chemical substances, guided by full consideration of reaction efficiency, can represent an effective method for decreasing the health and environmental impacts of chemical synthesis.

Chapter 9: Alternative Feedstocks and Products

In the preceding chapters, we have considered a variety of ways to decrease the health or environmental impacts of a chemical synthesis or process. In this chapter, we consider a more fundamental reengineering of chemical processes, either to utilize different starting materials or to prepare alternative, safer products.

9.1   Alternative Feedstocks

Many starting materials ("feedstocks") for the synthesis of organic compounds are derived from coal or petroleum. This is due in large part to the remarkably diverse array of organic compounds that may be obtained from these fossil fuel sources. As but one example, the high-temperature treatment of coal ("carbonization") affords a material known as coal tar. From coal tar may be isolated a range of organic compounds, including aromatic compounds (benzene, toluene, xylene, naphthalene, and others), hydroxylated aromatics (phenol, cresol, and others), and nitrogenous bases (pyridine, methylated pyridines, and quinoline), as well as various gases (methane, ethane, ethene, etc.). Perkin's synthesis of mauve dye – perhaps heralding the arrival of the modern age of chemical synthesis, as outlined in Chapter 1 – began with aniline, derived from the coal tar product benzene. Heavy reliance on coal- and petroleum-derived feedstocks continues to this day.

Unfortunately, there are several significant problems associated with these traditional feedstocks. Perhaps most obviously, coal and petroleum are nonrenewable resources, and although current supplies appear to be ample, it may be shortsighted to continue to depend on unlimited availability of these sources of chemical feedstocks. In addition, the extraction of coal and petroleum from the earth comes at some environmental cost, highlighted by the scars left by strip-mining or beaches befouled by oil spills. In addition, miners are exposed to considerable health and safety risks as well. Finally, the processing of coal and petroleum to produce chemical feedstocks can lead to the generation of undesirable byproducts, including hydrogen sulfide, carbon dioxide, and polycyclic aromatic hydrocarbons.

Development of alternative sources of starting materials represents a significant method for the reduction of the environmental impacts of chemical processes. Use of renewable resources, in particular plant-derived matter ("biomass"), is receiving a great deal of attention, and promising results are being obtained. While there is some reluctance to turn to biomass feedstocks, given their potentially seasonal nature (i.e., since they may not be available year-round) and the possibility of crop failures, time-tested crops such as corn and soybeans represent reasonably safe bets. Any risks must be offset against the opportunity to reduce reliance on imported oil and expand markets for domestic agriculture. Potentially lower costs and reduced waste generation are additional attractive features of biomass feedstocks.

The exploitation of biomass feedstocks provides an additional incentive, one that in fact led to the use of plant-derived materials in specialty synthesis long before issues of environmental impact became of concern. Many important organic compounds, particularly pharmaceutical products, are chiral, and often only one enantiomer displays the desired activity. The other enantiomer may be less active or inactive, but may also display unacceptable activity, as illustrated by the example of thalidomide, discussed in Chapter 2. Living organisms routinely produce and use chiral compounds, and thus biomass represents a promising source of a wide variety of starting materials for the synthesis of single enantiomers of chiral compounds. The so-called "chiral pool" has been exploited in this way for many years; examples of abundantly available chiral compounds are provided below.

amino acids
(threonine)

hydroxy acids
(tartaric acid)

carbohydrates
(galactose)

terpenes
(α-pinene)

alkaloids
(cinchonidine)

The "chiral pool" – examples of naturally occurring chiral molecules of potential utility in organic synthesis

Even biomass wastes can be of value. Thus, "corn steep liquor" serves as a fermentation medium for the production of penicillin, which in turn is used as a starting material for the preparation of "semi-synthetic" penicillin derivatives such as the well-known amoxicillin, as discussed in Chapter 7.

A comprehensive discussion of biomass feedstocks is beyond the scope of this text, but a series of examples should prove illustrative as to the potential power and versatility of this approach.

Applied CarboChemicals teamed up with Argonne National Laboratory and three other U. S. Department of Energy laboratories to develop a new cost-effective process to convert corn into commercial chemicals. By engineering a bacterium that can produce succinic acid by fermentation of corn-derived glucose, these researchers have opened the door for the synthesis of valuable compounds derivable from succinic acid, including 1,4-butanediol, tetrahydrofuran, N-methyl pyrrolidinone and other chemicals. These chemicals can be incorporated into polymers and solvents for use in clothing, fibers, paints, inks, food additives, and an array of other industrial and consumer products.

The Department of Energy Pacific Northwest National Laboratory has partnered with Biofine, a small Massachusetts company, in developing an economical method for turning paper mill waste into levulinic acid. Preparable from virtually any biomass waste product, levulinic acid in turn allows the preparation of other important chemicals, including an alternative fuel additive, methyltetrahydrofuran.

Scientists throughout the world are exploring the conversion of plant oils into "biodiesel," which may serve as a substitute for petroleum-derived diesel fuels.

Professor James Crivello, of the Rensselaer Polytechnic Institute, is developing novel plastics based on soybean oil. Coating objects with epoxidized soy oil, followed by polymerization, results in formation of lightweight yet tough and resilient polymer coatings.

Professor Wolfgang Glasser, of Virginia Tech University, explores the technological uses of natural (principally wood-derived) polymers and biocomposites, developing lignin-based polyurethane resins. Other scientists are preparing epoxies, acrylics, and other polymer types from lignin.

The dominance of synthetic fibers and dyes in the textile industry was perhaps best epitomized by the brightly-colored polyester leisure suit of the 1970's. Turning the synthetic tide, western societies are increasingly turning to natural textile fibers such as cotton, and even natural indigo dye, made virtually obsolete by the synthetic dye industry (Chapter 1), is again becoming a significant dyestuff.

Although biomass feedstocks offer considerable promise, it is important to maintain the same "big picture" perspective taken when exploring other green issues. Potential downsides to the use of biomass include the environmental costs associated with the growth of the biomass –water resource depletion for irrigation, application of fertilizers, pesticides, and herbicides, etc. – and waste generation in the production of materials from biomass feedstocks. Adding to the complexity of such an environmental cost analysis, biomass-derived textiles may require dry-cleaning instead of simple hand- or machine washing. (The full analysis of environmental impacts forms the basis of the discussions in Chapter 10.)

9.2   Alternative Products

Although some applications of organic chemicals – e.g., in the pharmaceutical industry – require the preparation of a particular molecular structure, many applications require only particular molecular *properties* rather than specific chemical structures. In these cases, rather than attempting to develop greener procedures for the preparation of a particular compound, it can be of great value to consider just what properties are required of the compound. By exploring the possibility of alternative structures that may display these desired properties while also displaying fewer hazards (in production or use) than the original compound, human health and/or environmental impacts can be reduced.

The preceding discussion of alternative feedstocks provides a number of representative examples of alternative product selection. For example, if one desires a mixture of organic compounds that will burn well enough to power an internal combustion engine, perhaps viable alternatives to gasoline such as "biodiesel" will serve this purpose. If one desires strong polymer coatings, perhaps plant oil-based resins will be as effective as petrochemically-derived polymers. If textile fiber production and dying is overly reliant on fossil fuel resources and environmentally destructive, perhaps one should consider the use of natural fibers and dyestuffs. It is important to note again, however, that each of these alternatives comes with its own potential environmental or health impacts when the "big picture" is considered (Chapter 10). However, "out of the box" thinking along these lines represents the opportunity for potentially major changes.

As in the preceding discussion of alternative feedstocks, while a comprehensive discussion of alternative products is beyond the scope of this text, several additional examples of alternative product design should prove illustrative.

Dow AgroSciences developed the Sentricon™ system for controlling subterranean termites. Replacing non-selective fumigants such as methyl bromide, which are hazardous to desirable organisms (and humans) as well as to pests, this system relies on the social nature of termites to distribute a selective poison. Foraging termites visit bait stations and return to the colony with food containing hexaflumuron, a chitin growth inhibitor that stops the molting process essential for termite growth.

Hexaflumuron, a designed selective termiticide

Carcinogenic dibromochloropropane is being replaced as a soil fumigant and nematocide with "solarization" processes using transparent polymer films to trap solar energy and raise soil temperatures [38].

In recent years, Rohm and Haas has developed CONFIRM™ (tebufenozide), an agent that selectively disrupts the molting process in destructive caterpillars while leaving other insects unharmed, and Sea-Nine™, a marine antifoulant that replaces tributyltin oxide. Widespread use of the latter had led to high levels of toxic tin compounds in seawater.

---

38.   J. Katan, in "Innovative Approaches to Plant Disease Control," I. Chet, Ed.; John Wiley & Sons: New York, 1987.

tebufenozide (CONFIRM™)

Sea-Nine™

Other chemical products designed to minimize environmental impacts

Ozone-damaging chlorofluorocarbons have been largely replaced with hydrofluorocarbons for use as propellants, refrigerants, etc.

These developments are most encouraging, and it is expected that alternative product design will continue to play an important role in the coming years. As always, however, it is important to raise a cautionary flag. After all, DDT was viewed as a miracle discovery, promising the eradication of insect-borne diseases, and was only later discovered to have unacceptable environmental impacts [39]. As new, greener products are engineered, careful attention must be given to potentially unforeseen hazards as well as to the full life cycle analysis of the replacement materials.

Summary

By thinking creatively and exploiting the wealth of knowledge we have gained about the structure and reactivity of organic molecules through the past 150 years, it is possible to select alternative starting materials (feedstocks) that reduce our dependence on nonrenewable resources like coal and petroleum. In cases where chemical properties are more important than specific chemical structures, greener alternative products may be designed to replace existing compounds that represent undesirable health or environmental risks.

---

39. This statement actually foreshadows the discussions in Chapter 10, which deals with the complexity of green chemical analysis. While DDT unarguably causes serious environmental harm, it is also very effective at eradicating insects that carry serious human diseases such as malaria.

Chapter 10: The Big Picture and Green Chemistry Metrics

In this final chapter before introduction of the experimental work, the complexity of an analysis of the "greenness" of a chemical reaction or process is examined. Relative rather than absolute measures of "greenness" are discussed, and economic considerations are added to the discussion.

10.1   The Challenge and Complexity of Implementing Green Chemistry

Through the discussions of the preceding chapters, it has presumably become quite obvious that green chemistry can be very complicated. Even seemingly simple choices – e.g., replacement of benzene as a solvent with water – become less simple when one examines the full ramifications of the choice. In this example, water is unarguably safer to work with than benzene. If water is used as a solvent, however, it is critical that it be purified before it is released to the environment, since any contaminants will easily be passed on to water-dwelling organisms and, in turn, to humans. Water is appreciably more difficult than benzene to purify once a reaction is completed, requiring at the least a higher energy input in order to distill it (bp 100 °C vs. 78 °C for benzene). This energy typically comes from burning fossil fuels, an activity that consumes an irreplaceable resource, contributes to global warming through CO_2 production, and releases a variety of other pollutants into the atmosphere, waterways, and soil.

A significant contributor to the complexity of such an analysis is the need to consider the full set of interrelated parameters – a "cradle to grave" or "life cycle" analysis is required for each reactant, reagent, solvent, product, and byproduct. Putting aside the question of whether or not a product should be made in the first place (!), a complete analysis of the greenness and feasibility of a reaction requires consideration of a remarkably complex set of parameters, some purely chemical and others more pragmatic. A number of criteria have been discussed in the preceding sections and are listed below.

- Relative chemical hazards of reagents
     Toxicity
     Environmental impact
     Reactivity, flammability, etc.

- Solvents/reagents required for reaction
- Solvents/reagents required for purification
- Efficiency of reaction
- Reaction time
- Reaction temperature
- Energy consumption
- Ease of separation
- Purity of product
- Hazards of byproducts
- Ease of recovery, reuse
- Ease of product isolation

Many of these criteria are more easily addressed by comparison of a proposed green method with an existing method. Other criteria for consideration – not specifically addressed in our earlier discussions, but of importance in a full analysis of a reaction or process – include the following:

- Environmental impact of manufacture of starting materials, reagents, and solvents – If a starting material, reagent, or solvent is only available at considerable environmental cost, whether it be through required extraction of petroleum resources, multi-step manufacture, or some other cost, it does not really represent a green advantage.

- Environmental ramifications of waste disposal – Although it is easy to focus on the volume of waste generated, an alternative process must also yield less hazardous waste. Which is worse, a process generating a ton of sodium chloride or a pound of 2,3,7,8-tetrachlorodibenzodioxin (TCDD)? What about a ton of benzene vs. a gram of TCDD?

- Transportation hazards – The required materials must be safely transportable. Ordinarily, the greener the starting material, reagent, or solvent, the lower the transportation hazard. However, recognizing that "green" is a relative term, this issue can arise. For example, a new process for the preparation of the analgesic, ibuprofen, utilizes substantial quantities of liquid hydrogen fluoride as a solvent. The hazards presented by transportation of this volatile and highly corrosive substance are rather extreme.

- Difficulty of synthetic method – How complex a set of operations is required to carry out the procedure? Could a technician be trained to carry it out, and can the necessary equipment be engineered?

- Reproducibility – Does the procedure work every time, or does it require the right magical incantation and phase of the moon for it to work?

- Scalability – Can the process be carried from the exploratory laboratory scale (perhaps a few grams) to the real production level (perhaps a few tons?) Many practical engineering issues arise here, including heat transfer and mixing.

- Success in loss prevention – Does the new process really reduce emissions or other unintentional releases?

- Ease of glassware or reactor cleaning – If the cleanup after completion of a process requires volatile, toxic, or environmentally harmful solvents or is prohibitive in terms of time or expense, the process is not as green as envisioned or will not be economically viable.

As the latter point illustrates, one must recognize that, for an alternative process to be adopted, it must be not only scientifically sound but also economically viable. Economic considerations, ultimately, will play key roles in the success or failure of a proposed green process. These concerns include the following:

- Cost of starting materials, reagents, and solvents

- Waste disposal costs

- Conversion costs (training, manufacturing plant retooling)

Given the complexity of a complete analysis of a new reaction or process, it is often difficult to make direct comparisons with existing procedures. Consider, for example, a partial analysis of one of the experiments presented in this text – the tungstate catalyzed oxidation of cyclohexene to adipic acid – in contrast with a commonly used oxidation method utilizing nitric acid [40].

Comparison parameter	Nitric acid oxidation of cyclohexanol	Phase transfer/tungstate catalysis of cyclohexene
Reagent hazards	high - toxic, corrosive, reacts violently	Neutral - PTC somewhat toxic
Impact of starting material manufacture	Not good	Not good
Cost of reagents (approximate)	$0.20/gram	$0.52/gram (no catalyst recycling) $0.17/gram (catalyst recycling)
Reaction efficiency	High	High
Reaction time	Short (< 1 hr.)	Longer (ca. 3 hrs)
Reaction temp (°C)	80-90	100
Ease of separation	Good	Good
Toxicity of byproducts	Nitrogen oxides ("NO_x") are toxic and greenhouse gases (adipic acid synthesis results in 10% of NO_x releases into atmosphere each year).	Unknown, low?
Product purity	Good	Good
Ease of recovery/reuse	Impossible?	Easy
Waste disposal	Low	Low
Conversion costs	None	High

Is it obvious which is the better procedure? How would you decide and/or persuade your supervisor that a change would be appropriate?

---

40.   It is important to note that such comparisons also depend upon the scale and environment in which the procedures are being carried out. This comparison is based upon a teaching lab-scale preparation.

While a full analysis of the greenness of a new reaction or process is difficult, this does not mean that we should not try. As we have noted, green chemistry can be a matter of taking small steps in the right direction.

As you continue your studies, through the experiments contained in this text, in other courses, or in the "real world," keep taking those small steps. Before you know it, they will have taken you far!

Summary

Analysis of the health and environmental impacts of chemical processes requires a full analysis not just of the chemical process itself, but of all inputs and outputs from "cradle to grave." Given this complexity, it would be rare to find an example of a reaction or process that epitomizes all the ideals of green chemistry. The effort to decrease the negative health and environmental impacts of chemical synthesis and processing is simplified by the realization that "green" is a relative term. Although a modification may not reach perfection with regards to green concepts, it may be recognized as "greener" than an existing procedure. Not only chemical, but also economic issues are relevant in a full analysis of a chemical process or reaction.

Green Organic Chemistry:

Laboratory Experiments

Chapter 11: Preface to the Experimental Section

11.1   Introductory Comments

It is through the practice of organic chemistry that its concepts truly come alive. The following section contains a set of laboratory exercises designed to provide practical experience with the techniques and concepts of organic chemistry. In addition, each of these experiments has been developed to illustrate one or more of the essential thought processes of green chemistry.

These experiments have been developed to parallel roughly the content of a typical undergraduate organic chemistry lecture course. In many such courses, introduction of "real" organic chemical reactions is slowed by the need to establish a firm understanding of organic structure, bonding, and nomenclature. Thus, some of the chemistry involved in the earlier experiments may not yet have been covered in your lecture course. (For example, Experiment 1 involves a reaction known as the aldol reaction, generally covered in discussions of carbonyl compounds.) While we recognize this, we have intentionally included such experiments, feeling that the value of carrying out a real chemical reaction, followed by analysis of the outcome (in Experiment 1, focusing on melting point issues), more than outweighs the timing of introduction of relevant lecture course material.

This text does not present separate discussions of the basic laboratory techniques such as melting point determination, recrystallization, chromatography, and spectroscopy. As such, it will presumably be used in tandem with a more traditional techniques-oriented laboratory text or with supplementary materials provided by your instructor. Each experiment suggests, by topic area, those sections of your companion text that should be reviewed prior to carrying out your laboratory investigations.

Each organic chemistry laboratory instructor will have his or her own ideas about practical issues such as laboratory safety, waste disposal, and reports. We have included a minimal set of materials relating to these issues in this text, but your instructor will be the ultimate authority in these regards. We have also included suggested pre- and post-lab written exercises, with the pre-lab work designed to prepare you for the experiment and alert you to any significant safety issues and the post-lab work designed to facilitate the analysis of the outcome of the experiment. Your instructor may wish to add to or replace these exercises with others of his or her choosing.

Many of these experiments lend themselves to inquiry-based exploration of alternative starting materials. Your instructor may advise you on the possibility of carrying out such explorations.

11.2 General Laboratory Safety Issues

1. Safety goggles must be worn at all times. Regular eyeglasses or contact lenses, by themselves, do not provide adequate protection.

2. Low-heeled shoes that cover the toe and instep must be worn at all times while in the laboratory. Clogs and sandals are not allowed.

3. It is a good idea to wear old clothes or a lab coat while doing lab work. Shorts are not allowed; pants that cover the ankles are required. Hair that is shoulder-length or longer should be put up or tied back.

4. Any student under the care of a physician for either acute or chronic medical reasons (e.g.; diabetes, hypertension, epilepsy, pulmonary disease) and/or is pregnant should advise their physician that they are taking organic chemistry lab and alert the laboratory instructor.

5. Proper planning is one of the keys to success and safety in this course. Make sure that you understand the procedures that you will use in the lab. If you don't know how to do something, ask.

6. Do not arrive at your lab section unprepared for the day's experiments. Ill-conceived lab work almost always leads to poor results and may be dangerous. Plan ahead!

7. Be prepared to respond to an emergency. Familiarize yourself with laboratory safety equipment, including eye wash stations, fire extinguishers, and fire blankets. Think about what you would do in the event of an emergency (fire, chemical spill, accident, etc.).

8. Don't use open flames unless specifically instructed to do so, and check carefully for the presence of flammable materials before lighting up.

9. Report all injuries, no matter how small.

11.3 Working With Chemicals

1. Know the hazards of <u>all</u> materials that you work with. When using a known compound, reagent or solvent, be familiar with its physical and toxicological properties (see Chapter 4).

2. If you are working with an unknown compound, be cautious. Treat all unknown materials as though they are potentially harmful.

3. Avoid ingestion of and/or contact with hazardous chemicals.
   - Don't handle compounds.
   - Never pipette by mouth.
   - Wear appropriate gloves when working with dangerous or unknown chemicals. Do not touch your face while you are wearing gloves.
   - Do not eat, chew gum, drink, or smoke in the laboratory.
   - Transfer toxic or offensive liquids in a fume hood.
   - Wash your hands and face thoroughly after finishing lab work for the day.

4. Do not ever mix chemicals together without knowing what you are doing. Mixing incompatible chemicals can result in violent reactions or release of toxic fumes that can harm you or others near you. If you are in doubt about the compatibility of reagents, ask.

11.4 Waste Disposal/Recycling

1. Place all waste in the appropriate waste or recycling container. Do not pour solvents, organic compounds, or strong acids or bases in the sink. If you are in doubt about how to dispose of something, ask.

2. Never pour any unused reagents back into stock bottles. You could contaminate the reagents and ruin your or others' experiments. Take only what you need, and be sure to replace caps after you have finished removing reagents.

3. Only paper should be discarded in the trashcans.

11.5 Sources of Chemical Safety Information

Chapter 4 provides a listing of a number of sources of information about the hazards presented by various chemicals. These sources should be consulted whenever faced with a new compound of unknown properties or with any question about handling or disposal issues.

11.6 Laboratory Notebooks and Reports

Your laboratory notebook is *the* place where you should record all your experimental details, observations, amounts, spectral results, and any other information relating to or resulting from your experimental work. In addition to recording experimental details, you should also answer all pre- and post-lab questions in your notebook. Make sure to provide clear and complete answers, neatly written and easily locatable on a separate page or pages (i.e., not hidden on a page covered with scrawled calculations).

A.   Pre-lab Reports

The following suggests a useful format for your pre-lab entries in your notebook

- *Title:* A suitable title for the week's experiment, in your own words.

- *Purpose:* A brief statement explaining the significance of the experiment and what you expect (or are expected) to learn from it.

- *Chemical Equation:* When appropriate (i.e., in experiments involving a chemical reaction rather than the extraction of a compound or the determination of a physical property), provide a balanced chemical equation.

- *Table of Physical Constants:* Include the names or structures of all starting materials, reagents, solvents, etc. that are to be used in the experiment. For each, provide the molecular formula, molecular weight, and quantity to be used (in grams and in moles). This table can also contain physical properties such as density, color, odor, etc. Include the source(s) of all data reported in this table. Include a listing of any significant health and safety issues, including those highlighted in the experimental procedures provided in this text.

- *Procedure:* A short handwritten procedure of how the experiment will be performed in the lab. For certain experiments, a flow chart will be both more informative and more appropriate and can substitute for the handwritten procedure. For this to be of any value to you, it must be in your own words – do not simply copy the procedure as reported in this text.

- *Answers to any Pre-Lab Questions.*

B.   Experimental Notebook Entries

It is important to record your observations and actual procedures in your laboratory notebook. If you have any doubts about proper notebook procedures, ask your instructor for advice on how to keep a better notebook. As a rule, think of your instructor while he/she tries to grade your notebook – ensure that everything is clear, complete, and legible.

C.   Post-lab Reports

The post-lab questions have been designed to allow you to present all the relevant data from your experiment, as well as to discuss some appropriate related concepts, within a concisely defined format.

D.   Economic Analysis

For each experiment, you should perform a basic economic analysis. This kind of analysis is relatively simple to carry out and can be very useful when comparing alternative synthetic methods, a sort of comparison that is essential to the development of and understanding of greener chemical reactions (c.f. Chapter 10). Your economic analysis may take the following form.

1. Identify the costs of all required materials.

    Use chemical catalogs such as that issued by the Aldrich Chemical Company to find the prices of your starting materials, reagents, and solvents. In many cases, you will find prices listed for a variety of quantities and/or purities of the desired compound. For your calculations, choose a level of purity that matches the material used in your laboratory work and use the least expensive option (usually the largest amount). It is worth noting that these prices/gram will be significantly higher than those a chemical company would pay for chemicals purchased "in bulk" (e.g., by the 55 gallon drum or tanker-load).

2. Record the total amounts of each material used.

    If you keep a good notebook, you will automatically record these amounts while you are working. If you use a material that is not typically measured precisely (chromatography solvents or drying agents, for example) make your best estimate of the amount used.

3. Calculate the cost for each starting material, reagent, or solvent and determine the total cost for the experiment.

Once you have determined the yield of your reaction, calculate the cost/gram of your product.

4. Monitor the amount of waste generated.

In each experiment where chemical waste is generated, keep track of the waste that you generated and how you disposed of the waste (e.g. liquid waste jug, solid waste, recycled). Assuming that it will cost $25/L to dispose of liquid waste and $25/kg to dispose of solid waste, show how the cost of waste disposal influences the cost of your product.

EXPERIMENT 1

SOLVENTLESS REACTIONS:
THE ALDOL REACTION

Chemical Concepts

Carbonyl chemistry; the aldol reaction; melting points of solids and mixtures; recrystallization.

Green Lessons

Solventless reactions between solids; atom economy.

Estimated Lab Time

1 – 2 hours

Introduction

The aldol condensation represents a powerful general method for the construction of carbon-carbon bonds, one of the central themes of synthetic organic chemistry. In the base-catalyzed aldol condensation reaction, deprotonation alpha (adjacent) to a carbonyl group affords a resonance-stabilized anion called an enolate, which then carries out nucleophilic attack at the carbonyl group of another molecule of the reactant. (Analogous acid-catalyzed reactions are also well-known.) The product, a beta-hydroxy carbonyl compound, often undergoes facile elimination of water (dehydration), affording an alpha, beta-unsaturated carbonyl compound as the final product.

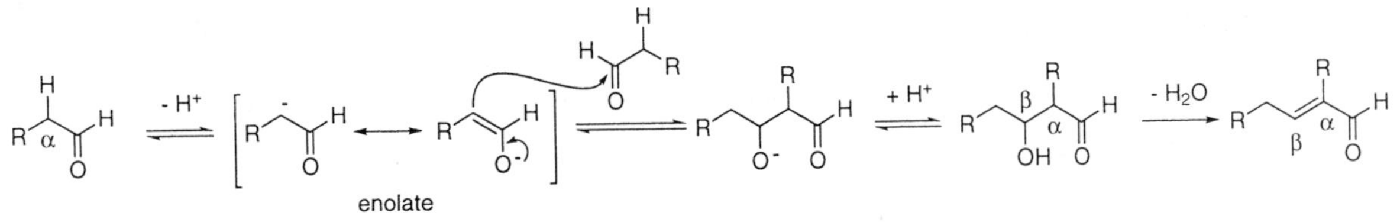

Mechanism of the base-catalyzed aldol condensation

Aldol condensation reactions between two different carbonyl compounds can lead to complex product mixtures, due to the possibility of enolate formation from either reactant and to the possibility of competing "homo" coupling rather than the desired "cross" coupling.

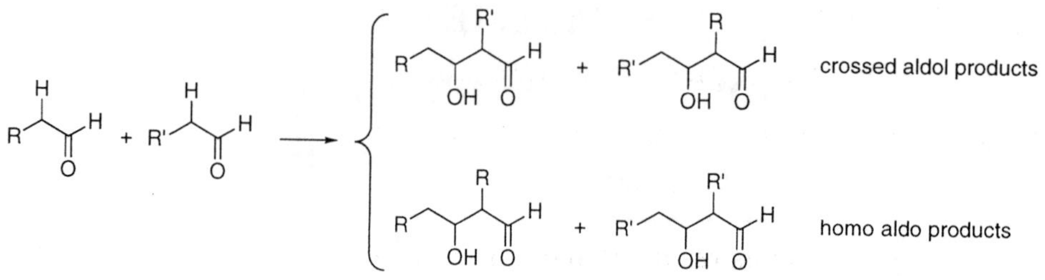

"Crossed" aldol condensation can afford complex mixtures

If, however, only one of the carbonyl compounds has alpha hydrogens available for deprotonation and enolate formation, the "crossed" aldol reaction can provide synthetically useful yields of products. Thus, for example, benzaldehyde cannot be converted to an enolate, yet reacts readily with enolates of other carbonyl compounds, including acetone.

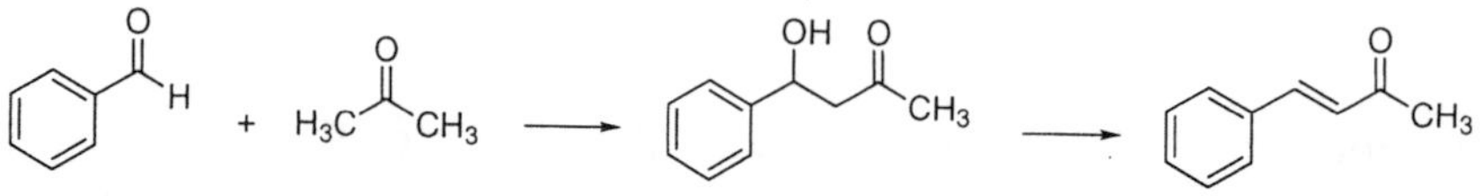

A successful "crossed" aldol condensation

Homo coupling of acetone (or other ketones) is generally not a problem in such reactions, as the aldol condensation of ketones is generally not a very efficient reaction. (More specifically, each step of the aldol condensation is reversible under the reaction conditions – at least until the dehydration step – and thus equilibrium is established. With aldehydes, equilibrium favors the aldol product, but with ketones, primarily for steric reasons, very little aldol condensation product is present at equilibrium.)

In this experiment, you will explore the aldol condensation reaction of 3,4-dimethoxybenzaldehyde and 1-indanone.

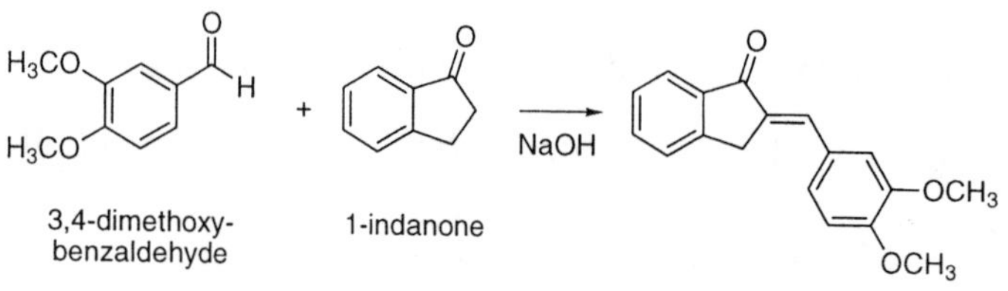

3,4-dimethoxy-
benzaldehyde

1-indanone

116

In contrast to typical experimental procedures for aldol condensation reactions, this reaction will be carried out without solvent. Ongoing research is revealing a number of reactions that proceed nicely in the absence of solvent, representing the best possible solution to choice of a benign solvent. Although these reactions are frequently referred to as "solid-state" reactions, it has been noted [41] that in many cases, mixture of the solid reactants results in melting, so that the reactions actually occur in the liquid, albeit solvent-free state. This melting phenomenon is interesting and actually represents one of the key points of this experiment. You have learned that impurities lead to lower melting points. Here, you will experience this in a vivid way – as you mix the two solid reactants, they will melt. In addition to providing a memorable demonstration of the impact of impurities on melting points and illustrating the possibility of carrying out organic reactions in the absence of solvents, this experiment highlights another key green concept – the design of efficient, atom-economical reactions. The aldol condensation, if effected without dehydration, has an atom economy of 100% and requires only a catalytic amount of acid or base, and even with dehydration, the atom economy remains quite high.

Pre-Lab Preparation

1. Study the technique sections in your lab manual regarding melting points and recrystallization.
2. Carry out pre-lab preparations as described in Chapter 11, section 11.6A, or as called for by your instructor.

Experimental Procedure

SAFETY PRECAUTIONS: Use care to avoid contact with solid sodium hydroxide or the reaction mixture.

41.   G. Rothenberg, A. P. Downie, C. L. Raston, and J. L. Scott, "Understanding Solid/Solid Organic Reactions," *J. Am. Chem. Soc.* **2001**, *123*, 8701-8708.

Reaction

1. Place 0.25 g of 3,4-dimethoxybenzaldehyde and 0.20 g of 1-indanone in a test tube. Using a metal spatula, scrape and crush the two solids together until they become a brown oil. Use care to avoid breaking the test tube.

2. Add 0.05 g of finely ground (using a mortar and pestle) solid NaOH to the reaction mixture and continue scraping until the mixture becomes solid.

Workup and purification

3. Allow the mixture to stand for 15 minutes, then add about 2 mL of 10% aqueous HCl solution. Scrape well in order to dislodge the product from the walls of the test tube. Check the pH of the solution to make sure it is acidic.

4. Isolate the solid product by vacuum filtration, continuing to pull air through the solid to facilitate drying. Determine the mass of the crude product.

5. Recrystallize the product from 90% ethanol/10% water, using the hot solvent first to rinse any remaining product from the test tube. You should not require more than 20 mL of solvent to effect this recrystallization.

Characterization

6. Determine the mass and melting point of the recrystallized product. (A typical melting point range is 178 – 181 °C.)

Post-Lab Questions and Exercises

1. Describe the physical properties (color and state) of your crude product. Report the mass and percent of theoretical yield of the crude product.

2. Report the color and melting point range of your recrystallized product. Report the mass and percent of theoretical yield of the recrystallized product.

3. Calculate the atom economy for the reaction.

4. Perform an economic analysis for the preparation of your product.

Experiment Development Notes

This experiment was adapted from the primary literature. Any number of solventless aldol reactions are possible [41], and it may be attractive to allow students some latitude in choosing their reactants, taking care to avoid unexpectedly hazardous reagents or products. The reactants reported here were chosen deliberately to highlight the melting point depression phenomenon; other pairs of reagents may or may not visibly melt upon mixing.

EXPERIMENT 2

BROMINATION OF AN ALKENE:

PREPARATION OF STILBENE DIBROMIDE

Chemical Concepts

Halogenation of alkenes; reactions at elevated temperature; vacuum filtration; melting point determination.

Green Lessons

Safer solvents; safer reagents.

Estimated Lab Time

2 – 2.5 hours

Introduction

Simple hydrocarbons are relatively unreactive. In order to enhance their reactivity, thereby allowing their elaboration into more complex molecules, it is often necessary to introduce more reactive functional groups. Alkenes (olefins) – hydrocarbons containing the carbon-carbon double bond functional group – may be "halogenated" to form alkyl halides, which are then capable of undergoing a variety of further chemical transformations.

The bromination of an alkene, in which bromine adds across the double bond to yield a "vicinal" (1,2-) dibromide, is an example of an addition reaction. The generally accepted pathway for this reaction involves an ionic mechanism in which the electron-rich alkene acts as a nucleophile (a species that attacks electron-deficient centers) and the bromine acts as an electrophile (a species that reacts with centers of greater electron density). As bromine and the alkene approach one another, the Br–Br bond becomes polarized ($Br^{\delta+}$-$Br^{\delta-}$). The more positively charged Br atom is transferred to the alkene, yielding a cyclic bromonium ion and a bromide ion. In a second step, a bimolecular nucleophilic substitution reaction (S_N2), the bromide ion attacks one of the carbon atoms of the of the cyclic bromonium ion, opening the three-membered ring and leading to the vicinal dibromide.

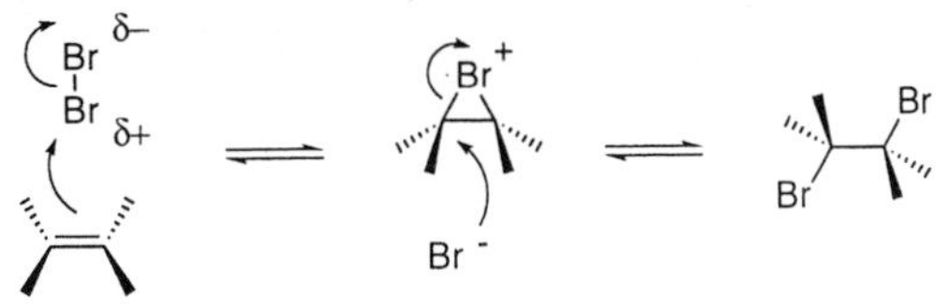

Mechanism for bromination of an alkene

In this experiment, you will carry out the bromination of (*E*)-stilbene (*trans*-stilbene), affording 1,2-dibromo-1,2-diphenylethane (stilbene dibromide). Typically, bromination of an alkene is accomplished using bromine in a chlorinated solvent such as tetrachloromethane (carbon tetrachloride) or dichloromethane (methylene chloride). Both of these solvents are suspected to be carcinogenic. Some brominations may also be carried out in glacial acetic acid, a volatile and corrosive liquid. We will use an alcohol, ethanol, as a safer alternative solvent.

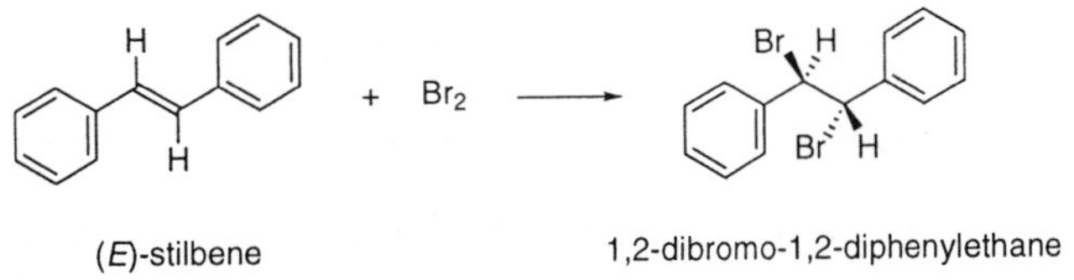

(*E*)-stilbene   1,2-dibromo-1,2-diphenylethane

Bromination of (E)-stilbene

Elemental bromine is volatile and highly corrosive, causing severe burns upon contact with the skin and extremely irritating upon inhalation. An alternative reagent, popularized by Djerassi and Scholz, is pyridinium tribromide [42]. This reagent exists in rapid equilibrium with pyridinium hydrobromide and bromine (reaction below) and thus provides for the "slow release" of bromine into the reaction

---

42.   C. Djerassi and C. R. Scholz, *J. Am. Chem. Soc.* **1948**, *70*, 417.

medium. (When a reagent is generated in the reaction medium rather than added to it, it is said that the reagent is prepared *in situ*.) An additional advantage of pyridinium tribromide is that it is an easily weighed solid. In contrast, liquid bromine is more difficult to measure out, by either weight or volume, due to its volatility, density, and hazardous and irritating vapors.

In situ generation of bromine from pyridinium tribromide

Pre-Lab Preparation

- Study the technique sections in your lab manual regarding gravity and vacuum filtration and melting point determination.

- Carry out pre-lab preparations as described in Chapter 11, section 11.6A, or as called for by your instructor.

Experimental Procedure

SAFETY PRECAUTIONS: Pyridinium tribromide is corrosive and a lachrymator. Avoid contact and clean up any spills immediately, particularly on the balance, the metal parts of which will quickly be corroded. Ethanol is volatile and flammable; avoid open flames.

Reaction

1. Place a magnetic stir bar, 2.0 g of (E)-stilbene, and 40 mL of ethanol in a 125 mL Erlenmeyer flask. (Be sure to record the exact mass of the stilbene you use.) Clamp the flask in place on a magnetic stirrer/hot plate. (The clamp will allow you to remove the flask from the hot plate without burning yourself in the event that the solution starts to boil too vigorously.)

2. With heating and stirring, dissolve the stilbene. (Be careful not to turn the heat up too far – the hot plate may be slow to heat at first, but then heat up very quickly. Once hot, it will take a long time to cool down again!)

2. Wearing disposable gloves, add 4.0 g of pyridinium tribromide. If solid material adheres to the interior walls of the flask, use a little ethanol to rinse it down.

3. Heat with stirring for 5 minutes after the addition of reagents is complete, then remove from the flask from the hot plate, using caution to avoid contact with the hot flask. The product dibromide should quickly begin to precipitate or crystallize.

Workup and Isolation

4. Let the reaction mixture cool to room temperature, then chill the mixture in an ice bath. Collect the product by vacuum filtration. Set aside a small amount of this "crude" product so that you can measure its melting point later.

5. Wash the isolated solid with a small amount of ice-cold methanol to remove any adsorbed pyridinium salts. Continue to draw air through your product until it is dry.

Characterization

6. Determine the mass of your purified product and measure its melting point. Also measure the melting point of the crude product. If time allows, record the melting point of (E)-stilbene in order to check the calibration of your melting point apparatus.

7. If time permits, confirm the identity of your compound by performing a mixture melting points determination with an authentic sample of stilbene dibromide [mp 241 °C (decomposition)]. You might also try a mixture melting point determination for a mixture of your product and (E)-stilbene.

Post-Lab Questions and Exercises

1. Describe the color and state of your purified product. Report the mass and percent of theoretical yield of the purified product.

2. Report the melting point *range* for your "purified" product and for the crude product. If you carried out any mixture melting point determinations, report the results.

3. Did the methanol rinse result in a more pure product?

4. Explain your evidence for your product's identity and purity.

5. In your words, explain the difference between melting and dissolving.

6. Bromide is a better nucleophile toward the intermediate bromonium ion than ethanol. What product would be obtained if ethanol did carry out this nucleophilic attack?

7. Calculate the atom economy for the reaction. How does it compare with the alternative bromination procedure described in Experiment 3?

8. Perform an economic analysis for the preparation of dibromostilbene via this route.

Experiment Development Notes

This experiment was developed based upon initial leads from Wilcox and Wilcox (*Experimental Organic Chemistry*, Prentice Hall, 1995) and the original report of the tribromide reagent by Djerassi and Scholz [42]. The reaction has been made greener by substituting ethanol for acetic acid as the reaction solvent in addition to substituting pyridinium hydrobromide perbromide for liquid bromine.

EXPERIMENT 3

A GREENER BROMINATION OF STILBENE

Chemical Concepts

Halogenation of alkenes; reactions at elevated temperature; reflux; vacuum filtration; melting point determination.

Green Lessons

Safer solvents; safer reagents; relative nature of green chemistry.

Estimated Lab Time

1 – 2 hours

Introduction

Experiment 2 provided an introduction to and discussion of the halogenation of alkenes. Since this experiment replaces hazardous solvents with a relatively innocuous solvent, ethanol, and replaces bromine with a safer and easier to handle reagent, pyridinium tribromide, it represents a relatively green procedure. However, Experiment 2 is not without its own limitations. As your calculations presumably revealed, it is not a particularly atom economical reaction. In addition, the preparation of the reagent, pyridinium tribromide, requires the use of elemental bromine. Thus, we have simply shifted the use of this hazardous reagent away from the bromination step rather than doing away with it entirely.

As advances in organic reaction chemistry continue to be made, further green improvements are made possible. In this experiment, we will take advantage of the *in situ* generation of bromine, through the oxidation of hydrobromic acid with hydrogen peroxide [43], to effect the bromination of (*E*)-stilbene to 1,2-dibromo-1,2-diphenylethane (dibromostilbene). (Refer to Experiment 2 for the chemical details

---

43. Rothenberg and Clark, "On Oxyhalogenation, Acids, and Non-mimics of Bromoperoxidase Enzymes," *Green Chemistry* **2000**, (2), 248-251.

of this transformation.) Although the requisite reagents – hydrobromic acid and hydrogen peroxide – must be handled with care, they appear to represent an improvement over pyridinium tribromide, as your post-lab analyses will indicate.

$$2\,HBr + H_2O_2 \longrightarrow Br_2 + 2\,H_2O$$

In situ formation of bromine by oxidation of hydrobromic acid with hydrogen peroxide

Pre-Lab Preparation

- Study the technique sections in your lab manual regarding heating a reaction at reflux, vacuum filtration, and melting point determination.
- Carry out pre-lab preparations as described in Chapter 11, section 11.6A, or as called for by your instructor.

Experimental Procedure

> SAFETY PRECAUTIONS: Hydrobromic acid is corrosive. Avoid contact and inhalation of vapors, and clean up any spills immediately. Hydrogen peroxide (30%) is a strong oxidizer and will readily damage clothing and body tissues, including skin. Ethanol is flammable; avoid open flames.

Reaction

1. Prepare a hot water bath in a crystallization dish on a stirrer/hot plate.
2. Place a magnetic stir bar, 0.5 g of (E)-stilbene, and 10 mL of ethanol in a 100 mL round-bottom flask. Fit the flask with a water-cooled reflux condenser.
3. Clamp the flask so that it may be heated and stirred in the hot water bath. Stir while heating the mixture to reflux, and continue heating and stirring until the majority of the solid has dissolved.

4. Slowly add 1.2 mL of concentrated aqueous hydrobromic acid. This will probably cause some of the stilbene to precipitate, but continued heating and stirring should cause the majority of the solid to redissolve. (Go on with the next step even if some remains undissolved.)

5. Measure out 0.8 mL of 30% hydrogen peroxide and add it dropwise to the reaction mixture. The initially colorless mixture will change in color to a dark golden-yellow.

6. Continue to stir and heat the reaction mixture at reflux until the yellow color fades and the mixture becomes a cloudy white. This typically takes roughly 20 minutes at reflux.

Workup and Isolation

7. Remove the flask from the hot water bath and allow it to cool to room temperature. Checking with pH paper, carefully adjust the pH of the solution to pH 5 to 7 through the addition of concentrated aqueous $NaHCO_3$. In some cases, very little $NaHCO_3$ is required.

8. Cool the reaction mixture in an ice bath to bring more product out of solution. Collect the solid that forms by vacuum filtration, rinsing with cold water. A wash with very cold ethanol can help to remove traces of impurities, but care must be used to avoid dissolving inordinate amounts of the product. Continue to draw air through your product until it is dry.

Characterization

9. Determine the mass of your product and measure its melting point [literature value: mp 241 °C (decomposition)].

Post-Lab Questions and Exercises

1. Describe the color and state of your purified product. Report the mass and percent of theoretical yield of the purified product.

2. Report the melting point *range* for your "purified" product.

3. Calculate the atom economy for the reaction. How does it compare with the alternative bromination procedure described in Experiment 2?

4. Perform an economic analysis for the preparation of dibromostilbene via this route.

Experiment Development Notes

Yields are typically around 90%. Replacement of the 30% hydrogen peroxide with household hydrogen peroxide (3%) is possible, but appears to lead to reduced yields (30-35%). We thank Professor Dieter Lenoir for the initial suggestion for exploring the *in situ* generation of bromine in this way. The experimental conditions represent adaptations from Olah, *et al.* [44] and Rothenberg & Clark [43].

44.   T.-L. Ho, B. G. B. Gupta, and G. A. Olah, "Phase Transfer Catalyst Promoted Halogenation of Alkenes with Hydrohalic Acid/Hydrogen Peroxide," *Synthesis*, **1977**, *10*, 676-677.

EXPERIMENT 4

PREPARATION AND DISTILLATION OF CYCLOHEXENE

Chemical Concepts

Dehydration of alcohols; multi-step syntheses; liquid-liquid extraction; drying agents; simple and fractional distillation; boiling point determination; infrared (IR) spectroscopy.

Green Lessons

Safer reagents; solvent-free synthesis.

Estimated Lab Time

3.5 hours

Introduction

Alkenes are frequently prepared by inducing the elimination of water (dehydration) from alcohols. The elimination reaction is typically induced by heating the alcohol with an acid catalyst.

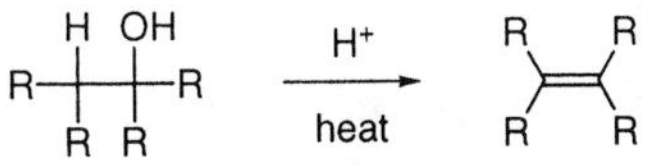

Dehydration of alcohols forms alkenes

In this experiment we will synthesize cyclohexene by the acid-catalyzed dehydration of cyclohexanol.

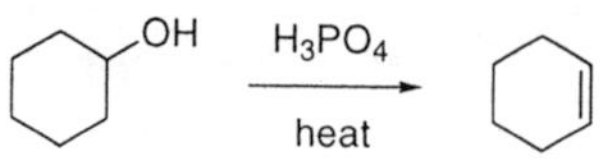

Preparation of cyclohexene by dehydration of cyclohexanol

The accepted mechanism for this reaction involves a three-step sequence, shown below. In the first step, the alcohol is protonated by the acid. This reaction is both rapid and reversible. In a second, slower step, loss of a molecule of water generates a carbocation. In the final step, a proton is lost from the carbon adjacent to the carbocation center, forming the alkene. Because the carbocation is a high-energy species, the ease of dehydration depends upon the structure of the carbocation. The more stable the intermediate carbocation, the more rapid the rate of dehydration.

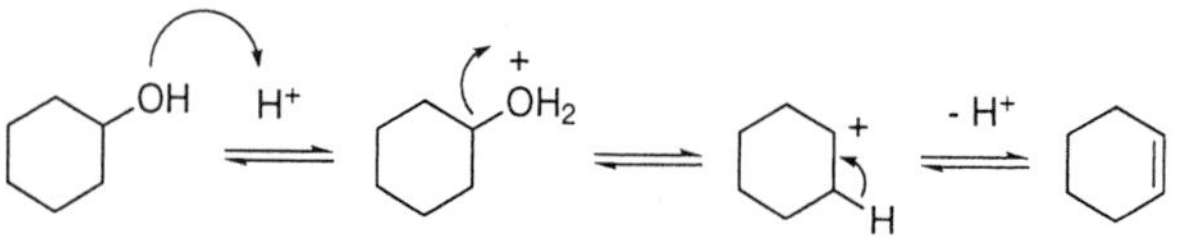

Mechanism for acid-catalyzed dehydration of an alcohol to an alkene

It is important to note that each step in this mechanism is reversible, and thus each species is in equilibrium with the others. Among other things, this means that the reaction itself is potentially reversible, and under certain conditions, one can indeed effect the acid catalyzed addition of water to an alkene, forming an alcohol.

Strong acids such as sulfuric acid are traditionally used to catalyze the dehydration of alcohols to alkenes. The use of sulfuric acid has two limitations, however. In addition to being highly corrosive, it can cause extensive charring of organic compounds due to its extreme reactivity. In this experiment, you will use a milder reagent, phosphoric acid, to effect the dehydration of cyclohexanol. Though phosphoric acid is still corrosive, it is much less reactive than sulfuric acid, thereby representing a reduced safety risk, and is far less likely to destroy your starting material, leading potentially to higher isolated product yields. In addition to introducing the use of a "greener" reagent, this experiment does not use any organic solvents. In this case, it is easy to eliminate the solvent because the reactants are all liquids that mix readily. Although elimination reactions are often not particularly green, since they can have relatively poor atom economy (see Chapter 8), in this case the elimination reaction leads to the formation of water, certainly one of the most innocuous of possible byproducts.

Pre-Lab Preparation

1. Study the technique sections in your lab manual regarding clamps, boiling, distillation, extraction, drying agents, and infrared spectroscopy.
2. Carry out pre-lab preparations as described in Chapter 11, section 11.6A, or as called for by your instructor.

Experimental Procedure

SAFETY PRECAUTIONS: Phosphoric acid, while safer than sulfuric acid, is corrosive. Avoid contact, and clean up any spills immediately. Cyclohexanol does not appear to present any unusual safety hazards. Cyclohexene is flammable and has a disagreeable odor.

Reaction

1. To a 50 mL round-bottom flask containing a magnetic stir bar (or boiling stone), add 0.074 moles of cyclohexanol and 1.75 mL of 85% H_3PO_4. Use gentle swirling to mix the two layers.
2. Fit the flask with a fractionating column, a distillation adapter, a thermocouple (or thermometer), a condenser, and a vacuum adapter as for fractional distillation (see illustration). A rubber septum should be used to provide a seal between the thermocouple or thermometer and the glassware. Be sure that the seal is good – if it is not, cyclohexene will escape from your glassware, causing your experiment to fail, and those of your classmates who find the odor of cyclohexene objectionable

will complain loudly! A drying tube, as shown in the illustration, can help to control the disagreeable odor of cyclohexene.

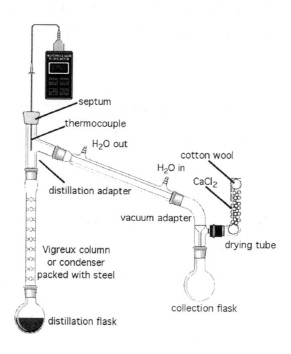

Apparatus for dehydration of cyclohexanol

3. Heat the reaction mixture first at a *gentle* reflux for about 5 minutes, then heat the flask more strongly in order to distill the mixture into the collection flask. Keep distilling until the volume remaining in the distillation flask has been reduced to approximately 1 mL.

Workup

4. Transfer the distillate to a separatory funnel and wash with approximately 5 mL of water. Carefully separate the layers and transfer the organic layer into a small, dry Erlenmeyer flask. If any water droplets are visible, remove them before adding the drying agent (sodium sulfate). Add a small amount of anhydrous sodium sulfate to the flask. Let the mixture stand for 5 minutes, occasionally swirling it gently. If the drying agent completely clumps together, its capacity to remove water has been exceeded and a little more sodium sulfate should be added. If you have successfully removed the water, the liquid should be clear, and at least a little of the drying agent should remain free flowing.

5. Decant or pipette the organic liquid away from the drying agent and place it in a clean, dry round-bottom flask. This will be the distillation flask for the next step. The appropriate size depends upon your yield. The flask should be about half full at the beginning of the distillation.

Distillation

6. Fit the flask with a distillation adapter and condenser in preparation for a simple distillation. (The apparatus will look the same as that used for fractional distillation, except that there will be no Vigreux or other fractionating column.

7. Be sure that your thermometer or thermocouple is properly positioned in order to measure the temperature of the distilling liquid accurately (see the illustration above). Carefully distill the organic material, collecting the material that distills in the range of 80 – 90°C. Typically there will be very little material remaining in the distillation flask. Be sure to record the boiling range that you observe.

Characterization

8. Transfer the distilled cyclohexene to a clean, dry, pre-weighed sample vial and determine the mass of the product. If time permits record an infrared spectrum of the distilled product.

Storage

9. You will need the cyclohexene for Experiment 5 – Synthesis of Adipic Acid. Keep it in a well-sealed and suitably labeled sample vial until then.

Post-Lab Questions and Exercises

1. Describe the color and state of your purified product. Report the mass and percent of theoretical yield of the purified product.

2. What boiling point range did you observe during your (a) initial distillation and (b) your final distillation? How do you explain the difference between these, if there was one?

3. What role does phosphoric acid play in this reaction? Use both words and chemical equations to answer this question.

4. Attach your infrared spectrum (if you obtained one) and identify ("assign") the major peaks in the spectrum. If you did not have time to obtain your own spectrum, assign the major peaks in a spectrum obtained from your instructor.

5. Calculate the atom economy for the reaction.

6. Perform an economic analysis for the preparation of cyclohexene via this route.

7. Predict the dehydration product(s) for the following alcohols. Show all anticipated products.

Experiment Development Notes

This experiment is presented in many organic lab texts in various forms. Some use sulfuric acid as a catalyst; others call for the use of phosphoric acid. We use phosphoric acid as a greener reagent. This dehydration chemistry may be readily extended to other alcohols, as hinted at in post-lab exercise 7, and this offers an ideal opportunity for inquiry-driven investigation. (Naturally occurring terpene alcohols are particularly attractive substrates.) In many cases, mixtures of isomers may be anticipated, calling for the use of additional analytical tools to determine the outcome of the reaction. A later edition of this text will include experimental details for the dehydration of α-terpineol and gas chromatographic analysis of the resulting product mixture.

EXPERIMENT 5

SYNTHESIS AND RECRYSTALLIZATION OF ADIPIC ACID

Chemical Concepts

Oxidative cleavage of an alkene C=C bond; phase transfer catalysis; recrystallization; melting-point determination; polymer chemistry.

Green Lessons

Catalysis; alternative reaction media; reuse of reagents.

Estimated Lab Time

4 hours

Introduction

The carbon-carbon double bond of an alkene, representing a site of relatively high electron density, is susceptible to oxidation. Depending on the reagent(s) and conditions used to effect the oxidation of an alkene, a variety of products may be obtained. With relatively mild oxidation, only the pi bond of the alkene is cleaved, affording, for example, epoxides or 1,2-diols. More vigorous oxidation can result in the complete cleavage of the carbon-carbon double bond, leading to the formation of various carbonyl compounds, with the specific product(s) dependent on the substitution pattern of the alkene and on the nature of the oxidant used.

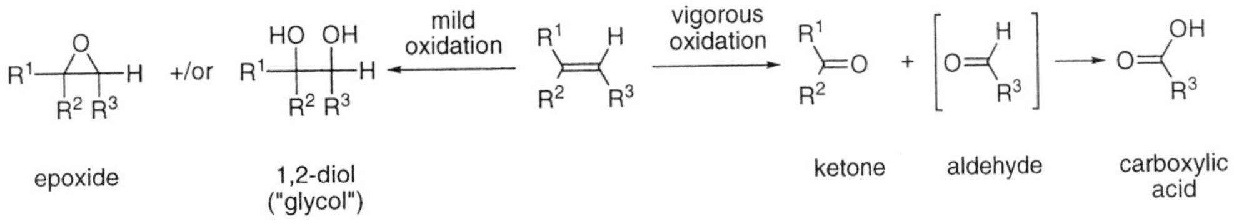

Representative products resulting from the oxidation of alkenes

The oxidative cleavage of alkenes, in many cases, is believed to proceed through initial oxidation of the double bond to a 1,2-diol (a "glycol"). Further oxidation of this intermediate 1,2-diol results in cleavage of the carbon-carbon bond. If the cleavage product is a ketone, the oxidation generally stops at this stage, but if an aldehyde is produced, it will usually be further oxidized to the corresponding carboxylic acid, as shown above. The precise details of this type of reaction (including the identity of all of the intermediates along the reaction pathway) are not fully understood.

In this experiment, we will carry out the oxidative cleavage of a simple cyclic alkene, cyclohexene, yielding 1,6-hexanedioic acid (adipic acid).

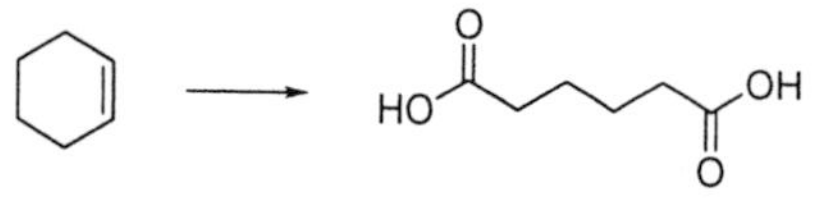

Synthesis of adipic acid via oxidative cleavage of cyclohexene

Adipic acid is used in the production of "Nylon 6,6," comprised of alternating units of adipic acid and 1,6-diaminohexane joined through amide linkages. Other nylons may be prepared by forming amides from other dicarboxylic acids and diamines. The nylons are among the most important of the commercial "polymers," large molecules made up of many covalently-linked smaller molecules ("monomers"). A monomer is a molecular building block that can be used to make up a polymer chain when bound together with other monomers.

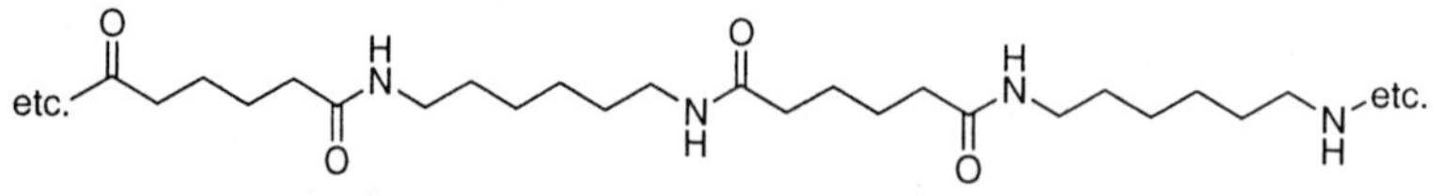

Nylon 6,6 – a polymer of adipic acid and 1,6-diaminohexane

The laboratory-scale oxidative cleavage of alkenes is typically accomplished with hot, basic potassium permanganate ($KMnO_4$) solution. This method involves a harsh oxidizer and produces large quantities of MnO_2 waste. The traditional industrial synthesis of adipic acid uses nitric acid, a strong oxidizing acid that presents many chemical safety hazards and environmental risks. Nitric acid can react violently with organic compounds, and this has resulted in a number of serious accidents. Additionally, the use of nitric acid in the preparation of adipic acid results in the emission of nitrous oxide (N_2O), a suspected green house gas, and adipic acid production is believed to be the source of roughly 10% of all non-natural nitrogen oxide ("NO_x") emissions.

In this experiment, we will explore the use of an alternative oxidation procedure, using sodium tungstate (Na_2WO_4) as a catalyst for the oxidation of cyclohexene to adipic acid by hydrogen peroxide [45]. This hydrogen peroxide procedure is greener than the nitric acid reaction and is also appreciably greener than the traditional permanganate reaction, avoiding the need for strongly basic reaction medium and generating only water as a byproduct rather than MnO_2. Although the mechanism of this reaction has not been established, tungstate presumably acts as the active oxidizing agent, given its structural similarity to permanganate. Significantly, and illustrative of another green feature of this reaction, the tungstate functions catalytically, with intermediate reduced tungsten products being oxidized back to the tungstate oxidation state by hydrogen peroxide.

Sodium tungstate displays appreciable solubility only in water, while cyclohexene is virtually insoluble in water. Thus, when cyclohexene and an aqueous solution of sodium tungstate and hydrogen peroxide are mixed, there will be two immiscible liquid phases present. To keep this reaction as green as possible, we want to avoid the use of any other solvents. In order to effect the reaction between cyclohexene and tungstate, then, we need to enhance the solubility of tungstate in cyclohexene. This can be effected by the technique of "phase transfer catalysis." Ammonium salts bearing hydrophobic groups, while still ionic, are frequently soluble in media of appreciable less polarity and hydrogen-bonding capability than water. Through ion pairing, such salts can bring negatively charged species into these less-polar media, thereby making them available for reaction. This phenomenon, illustrated below, is referred to as phase transfer, since it effects the transfer of a reactant from one phase to

45.   K. Sato, M. Aoki, and R. Noyori, "A 'Green' Route to Adipic Acid: Direct Oxidation of Cyclohexenes with 30 Percent Hydroxide Peroxide," *Science* **1998**, *281*, 1646.

another (here, from one immiscible liquid phase to another). Since the ammonium salt can play its role many times, it may be used in catalytic quantities, hence the name phase transfer catalysis for the overall process. In this experiment, we will use a commercial phase transfer catalyst known as Aliquat 336 [$(CH_3CH_2CH_2CH_2CH_2CH_2CH_2CH_2)_3NCH_3^+ Cl^-$]. Since both it and the sodium tungstate function catalytically, only hydrogen peroxide is needed as a stoichiometric reagent. In addition, although we won't do so in this experiment, it is possible to reuse the aqueous layer containing the tungstate catalyst, further enhancing the green nature of this procedure.

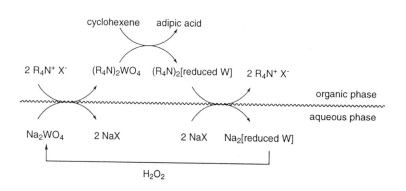

Phase transfer catalysis

Pre-Lab Preparation

1. Read the *Science* article describing the reaction method. While this paper is from the "primary" scientific literature, it is very readable and provides interesting background on environmental issues. In addition, it should provide you with some of the details necessary to complete this experiment.

2. Study the technique sections in your lab manual regarding heating a reaction at reflux, decantation, vacuum filtration, melting point determination, and recrystallization.

3. Carry out pre-lab preparations as described in Chapter 11, section 11.6A, or as called for by your instructor. Most of the required information can be obtained from the *Science* article, but note the following hints and modifications. 1) The 30% aqueous hydrogen peroxide solution is 30% *by weight*. 2) You will prepare your own phase transfer catalyst (a mixture of Aliquat 336 and

KHSO$_4$) rather than using the one described in the *Science* article. 3) Your molar ratios will deviate slightly from those reported in the *Science* article – use the quantities reported below.

Experimental Procedure

> SAFETY PRECAUTIONS: Avoid contact with the phase transfer catalyst as it can be irritating and can transport contaminants through the skin. Avoid getting hydrogen peroxide on you or your clothing. Cyclohexene is flammable, and its odor is often considered to be disagreeable.

Note: You need at least 2 g of cyclohexene to perform this experiment. If you did not obtain this quantity of cyclohexene from Experiment 4 (or if you did not carry out Experiment 4), obtain the necessary amount of cyclohexene from your instructor.

Reaction

1. Place 0.50 g of sodium tungstate dihydrate (Na$_2$WO$_4$•2H$_2$O) in a 50 mL round-bottom flask containing a stir bar and fitted with a water-cooled condenser. **Notes: The efficient stirring important for the success of this reaction is more easily achieved in a round-bottom flask than in a pear-shaped flask. An efficient water-cooled condenser is required to avoid loss of cyclohexene during the reaction.**

2. Add 0.5 g of Aliquat 336 – this is a very viscous liquid that is hard to transfer, so weigh it directly into your reaction flask. It is not necessary to obtain exactly 0.50 g. Next, add 11.98 g of 30% hydrogen peroxide and 0.37 g of KHSO$_4$ to the reaction mixture. Stir, then add 2.00 g of cyclohexene. **Note: The order of addition of the reagents is important.**

3. Heat the mixture to reflux on a sand bath, then continue to heat at reflux for 1 hour while stirring *vigorously*. About halfway through the reflux period, rinse down any cyclohexene trapped in the condenser with a few mL of water, added via pipette. Phase transfer catalysis depends upon very efficient mixing of the organic and aqueous layers, so it is important to stir as fast as possible throughout the reaction. (Generally, when using a magnetic stirrer, the closer your flask is to the

surface of the stirrer, the easier it is to maintain rapid stirring.) The reaction will not proceed if it is heated at *below* reflux, but it is also very important that you do not heat the mixture too strongly. If you do, cyclohexene may be lost through the top of the condenser. Watch the condenser closely – if you see liquid condensing near the top, you need to reduce the heat. You may need to remove the flask from the heat source temporarily in order to bring the reflux back under control. Stop the stirring occasionally to see if there are still two layers present. The reaction is complete when it no longer separates into two layers.

Workup

4. Use a pipette to transfer the hot reaction mixture into a small beaker, leaving behind any of the phase transfer catalyst that may have separated. (If the catalyst separates – and it does not always do so – it will stick to the walls of the flask or form a separate oily layer at the bottom of the flask. Careful execution of this step is the key to a successful purification. It is better to leave a little of the aqueous solution behind than to risk contamination of your solution with the phase transfer catalyst.)

5. Cool the beaker containing the reaction mixture rapidly in an ice bath. A precipitate should form within 20 minutes. Collect the crude product by vacuum filtration using a Büchner funnel.

6. After the crude material has air-dried, weigh it and determine its melting point.

Purification and Characterization

7. Recrystallize the crude product from the minimum required amount of hot water. Determine the mass and melting point of the recrystallized product and, (if time permits) obtain its infrared spectrum.

Post-Lab Questions and Exercises

1. Describe the color and melting point range of your crude product. Report the mass and percent of theoretical yield of the crude product.

2. Describe the color and melting point range of your recrystallized product. Report the mass and percent of theoretical yield of the recrystallized product.

3. What impurities might you plausibly expect to be in the crude product?

4. When you finished refluxing the reaction, did your reaction mixture still contain cyclohexene? (Were there two liquid layers in the reaction flask?)

5. Attach your infrared spectrum (if you obtained one) and identify ("assign") the major peaks in the spectrum. If you did not have time to obtain your own spectrum, assign the major peaks in a spectrum obtained from your instructor.

6. Calculate the atom economy for the reaction.

7. Perform an economic analysis for this preparation of adipic acid.

8. The commercial synthesis of adipic acid involves the oxidation of cyclohexanol with nitric acid. List the hazards involved with this process, and briefly comment on both the environmental impact and personal exposure issues associated with this commercial process.

Experiment Development Notes

The original literature report of this general procedure required extensive modification in order to make it suitable for use in the instructional organic chemistry laboratory. Most notably, the reaction time was decreased from 24 hours to roughly 3 hours, and an *in situ* method for generating the phase transfer catalyst was developed. This modified procedure has been published in the *Journal of Chemical Education* [46].

---

46. S. M. Reed and J. E. Hutchison, "An Environmentally Benign Synthesis of Adipic Acid," *J. Chem. Ed.* **2000**, 77, 1627-8.

EXPERIMENT 6

OXIDATIVE COUPLING OF ALKYNES:

THE GLASER-EGLINTON-HAY COUPLING

Chemical Concepts

Oxidative coupling of alkynes; decolorization; thin-layer chromatography (TLC); infrared (IR) spectroscopy; preparation of KBr pellets.

Green Lessons

Catalysis; alternative solvents; mild reagents (molecular oxygen).

Estimated Lab Time

4 hours

Introduction

In the preceding experiments, we have examined the transformation of one functional group to another – alkene to dibromide, alcohol to alkene, and alkene to carboxylic acid. While very important, these functional group interconversions do not provide for significant increases in molecular size or complexity. In order to synthesize more complex molecules, we must turn our attention to the construction of the carbon "skeleton" that represents the structural essence of any organic molecule. In this light, the formation of carbon-carbon bonds is one of the most important classes of reactions in organic synthesis. In this experiment, we will form a new carbon-carbon bond via the oxidative coupling of an alkyne.

$$R-C\equiv C-H \ + \ H-C\equiv C-R \ \longrightarrow \ R-C\equiv C-C\equiv C-R$$

Carbon-carbon bond formation: Oxidative coupling of alkynes

This coupling reaction requires oxidation rather than simple addition and elimination reactions such as those we have explored in the preceding experiments. You will use a transition metal (Cu) catalyst to effect this coupling. A catalyst lowers the activation barrier for a reaction, usually through some change in the mechanism, and participates in the reaction without being consumed, allowing it to effect many reactions. In this particular experiment, the copper catalyst also plays the role of a "template," facilitating the coupling reaction by bringing the two alkynes into close proximity.

In this experiment, you will explore the oxidative coupling of 1-ethynylcyclohexanol in the presence of cuprous chloride, tetramethylethylenediamine (TMEDA) and air, as shown below. This procedure, known as the Glaser-Eglinton-Hay reaction [47], has been employed in the synthesis of a number of fungal antibiotics. (As drugs, alkynyl compounds are often found to be more active, less toxic and more bioavailable than analogous alkanes or alkenes.)

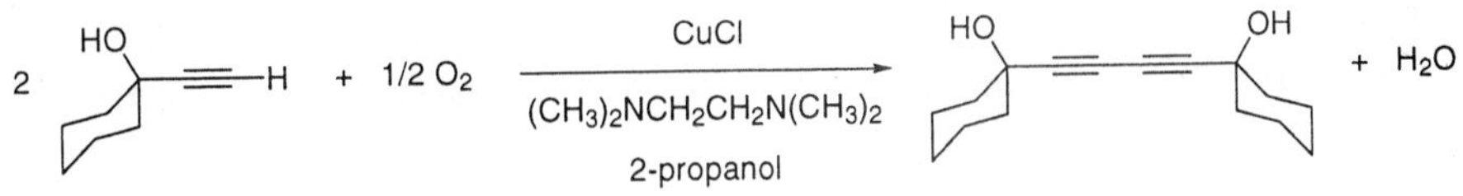

The Glaser-Eglinton-Hay coupling of 1-ethynylcyclohexanol

The mechanism of this reaction is fairly complex. Terminal alkynes are relatively acidic, with pK_a values of roughly 25 (compare the pK_a values of around 50 for simple alkanes), and the reaction is believed to commence with the deprotonation of the alkyne to form the corresponding acetylide. Subsequent reaction with Cu^{1+} affords a copper acetylide complex (believed to be a dimer, containing two cuprous ions and two acetylides). Oxidation of this complex by Cu^{2+}, formed by *in situ* oxidation of Cu^{1+} by air (oxygen), then leads to the alkyne coupling product.

The Glaser-Eglinton-Hay coupling reaction traditionally employs pyridine as both a reagent and as the solvent. Pyridine not only dissolves the alkynyl starting material, but also serves as a base to deprotonate the alkyne and as a "ligand," coordinating with the copper ion. However, pyridine has a number of adverse health effects, and in addition has a very objectionable odor that is detectable even at the 1 part-per-million level. In this experiment, a greener solvent, 2-propanol, is used in place of pyridine. Since 2-propanol is not appreciably basic and does not serve as a good ligand for copper, TMEDA is added to play these roles. TMEDA is safer than pyridine and, because it is not the solvent, can be used in much smaller quantities. Oxidation of Cu^{1+} to Cu^{2+}, required for this reaction to proceed catalytically, is safely effected by oxygen in the air.

In addition to alternative solvent selection, this experiment illustrates several more general green issues. Molecular oxygen (air), used as an oxidant, represents the epitome of green reagents – it is nontoxic and nonpolluting, requires no purification or special handling, and is even free – while exploitation of catalysis provides the advantages discussed in section 7.3E. Thus, catalysis can eliminate the need for high reaction temperatures or overly reactive reagents, helping to prevent product decomposition and byproduct formation. This translates, in turn, into reduced requirements for solvents, chromatography supports, or other materials for the isolation and purification of the desired product.

Pre-Lab Preparation

1. Study the technique sections in your lab manual regarding thin layer chromatography, decolorizing carbon, hot filtration, and solid sample preparation for IR spectroscopy.
2. Carry out pre-lab preparations as described in Chapter 11, section 11.6A, or as called for by your instructor.
3. What value do you expect for the melting point of the desired product?

47.   A. S. Hay, "Oxidative Coupling of Acetylenes. II," *J. Org. Chem.* **1962,** *27,* 3320.

Experimental Procedure

> SAFETY PRECAUTIONS: 2-Propanol and ethyl acetate are flammable; avoid open flames.

Reaction

1. To a 100 mL round-bottom two-neck flask containing a magnetic stir bar, add 30 mL of 2-propanol, 100 mg of cuprous chloride, and 20 drops of tetramethylethylenediamine. Stir with a magnetic stir bar while you assemble the rest of the apparatus – select one of the two setups shown below. (A third alternative, more complex to set up but equally effective, is presented at the end of this experiment.)

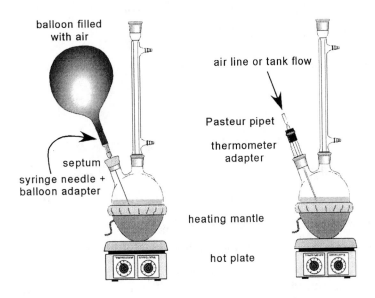

Apparatus for Glaser-Eglinton-Hay coupling of 1-ethynylcyclohexanol

Notes

i. The needle (or pipette) must be positioned so that its tip is below the solution surface.

ii. Maintain a slow, steady stream of air into the apparatus.

iii. If a three-neck flask is used, seal the extra neck with a septum – this neck can be used to add the alkyne.

iv. You may need to refill the balloon from time to time.

2. Add 2 g of 1-ethynylcyclohexanol to the reaction flask, then heat the reaction to a gentle reflux.

3. After about 30 minutes, spot a sample of the reaction mixture on a silica gel TLC plate. Spot the starting material in a separate lane on the same plate for comparison. Elute with a 70:30 mixture of hexanes/ethyl acetate, then use both ultraviolet light and iodine staining to visualize your TLC plate. If your reaction mixture still contains starting material, continue to heat the reaction at reflux until your TLC analysis shows that the reaction is complete. Some guidelines for carrying out a successful TLC analysis that may be useful to you are provided at the end of this experiment.

Workup

4. When the reaction is complete, evaporate the 2-propanol on a rotary evaporator. Ensure that *all* the 2-propanol has been removed – gentle warming will help to drive off the last traces.

5. Add 20 mL of water containing 1 mL of 12 *M* HCl to the material remaining in the flask. Collect the solid by vacuum filtration and pull air through it until it is dry.

Purification and Characterization

6. If your crude product is still blue or green, repeat step 5 to remove the remaining copper salts. If it is brown or black in color, decolorize it according to procedure 6A. If it is white or off-white, go directly to recrystallization procedure 6B. **Note: The odor of ethyl acetate is not unpleasant in small concentrations, but larger amounts can be troubling. Follow your instructor's recommendation as to whether to carry out these procedures in a fume hood or elsewhere.**

    6A. Dissolve the crude product in about 20 mL of ethyl acetate and add approximately 0.25 g of decolorizing carbon [48]. Heat the mixture gently for about 1 minute, then remove the carbon by filtration. Evaporate some of the solvent on a rotary evaporator. It should **not** be necessary to remove all of the solvent in order to cause the white crystalline product to separate.

    6B. Recrystallize the crude product by dissolving it in less than 20 mL of hot ethyl acetate, then allowing the solution to cool to room temperature.

---

48. Do not use more carbon than the amount specified – it will adsorb some of your product and decrease your yield!

7. Collect the product by vacuum filtration and pull air through it until it is dry. (If time allows, cool the solution in an ice bath for 10 minutes or so before isolating the product.)

8. Determine the mass of your product. Analyze the purity of your product by TLC. If time allows, record an infrared spectrum of the product as a KBr pellet. (Helpful tips for the successful preparation of a KBr pellet are provided at the end of this experiment.)

Post-Lab Questions and Exercises

1. Describe the color and melting point range of your product. Report the mass and percent of theoretical yield of the product.

2. In your own words, describe the roles of each of the reagents used during this reaction.

3. Describe your TLC results for both the reaction mixture and the purified product. How many spots did you observe in each case? Was there any difference between the two visualization methods (ultraviolet vs. iodine)? What were the R_f values for the spot(s)?

4. Attach your infrared spectrum (if you obtained one) and identify ("assign") the major peaks in the spectrum. If you did not have time to obtain your own spectrum, assign the major peaks in a spectrum obtained from your instructor.

5. Calculate the atom economy for the reaction.

6. Perform an economic analysis for the preparation of the product.

Experiment Development Notes

This experiment was patterned after experiments presented in Wilcox and Wilcox (*Experimental Organic Chemistry*, Prentice Hall, 1995) and Zanger and McKee (*Small Scale Synthesis: A Laboratory Textbook of Organic Chemistry*, W.C. Brown Co., 1995). These reported procedures were substantially modified, most notably by changing the solvent and glassware setup (enhancing the reliability of the experiment in our students' hands) and adding TLC training to the experiment.

Use of Negative Pressure to Supply Air

In the event that the specified apparatus is unavailable, the reaction can be performed using negative rather than positive pressure. In order to do so, the following modifications should be made.

i. Use a single-neck flask with a water bath for heating (maintain the water bath temperature at roughly 50 – 60°C.

ii. Attach a Claisen adapter. Insert a Pasteur pipette into a thermometer adapter and place this assembly into the straight neck of the Claisen adapter. The tip of the pipette must be below the surface of the solution.

iii. Place a straight vacuum adapter into the second neck of the Claisen adapter. Attach the side port to a vacuum source, and attach a separatory funnel containing roughly 30 mL of 2-propanol to the top of the vacuum adapter.

iv. Draw a *slow* vacuum on the apparatus, causing air to bubble slowly through the solution. If this does not happen, ensure that all of the glass joints are well sealed.

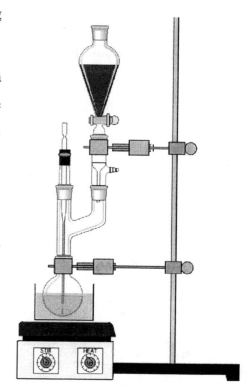

v. In the course of the reaction, some 2-propanol will be lost by evaporation from your reaction mixture. As needed, add 2-propanol from the separatory funnel to compensate for this loss.

A Short Guide to Successful TLC Analysis

i.   Apply your sample to the TLC plate

Using a glass capillary, apply a *small* spot of your reaction mixture (or compound dissolved in a suitable solvent) near one end of a TLC plate. The more concentrated your sample, the smaller the spot should be. If your solution is relatively dilute, build up a sufficient amount of compound for analysis by repeatedly spotting the sample at the same place on the TLC plate, waiting for the solvent to evaporate after applying each portion. If your capillary contains too much sample, it can be difficult to control the size of the spot – simply bleeding some of the solution out of the capillary by very briefly touching it to a piece of tissue before spotting the plate generally helps to control spot size.

Allow the spotting solvent to evaporate completely. If any of this solvent is left behind during elution, it may result in irreproducible changes in R_f values (the distance traveled by the substance divided by the distance traveled by the solvent front).

ii.   *Develop the TLC plate*

Place the TLC plate (with the sample spot down) in a developing chamber with a small amount of solvent in the bottom. Make sure the solvent does not cover the sample spot.

If possible, keep the atmosphere in the developing chamber saturated with the vapor of the elution solvent. (This may be easily accomplished by covering the chamber – a watch glass makes an ideal lid.) A strip of filter paper wrapped partially around the inside of the developing chamber is helpful, but not necessary. If used, ensure your TLC plate does not touch the filter paper, as this can cause solvent to bleed in from the edges of the plate, causing spots to drift diagonally across the TLC plate.

Allow the solvent front to approach, but not reach, the top of the TLC plate.

Remove the plate from the developing chamber, mark the solvent front with a soft pencil or a small scratch, and let the plate dry.

iii.   *Visualize the spots and calculate the R_f value*

Although some compounds may be visible to the naked eye on a TLC plate, many others are not. Most TLC plates contain ultraviolet light-sensitive phosphors, allowing use of an ultraviolet lamp for visualization. Alternatively, the plate may be placed in a capped bottle containing solid iodine; the iodine vapors will react with many compounds on TLC plates and cause them to become visible to the naked eye. Many other visualization methods are also available.

Mark the observed spots with a soft pencil and calculate their R_f values.

Tips for the Successful Preparation of KBr Pellets for Infrared Analysis

Thoroughly grind 1 – 2 parts of the compound with approximately 100 parts of dry, infrared-grade KBr with a small mortar and pestle. Do not make the common mistake of using too much compound! Transfer some of the mixture to the pellet press and apply pressure. The resulting pellet should be fairly transparent and uniform in thickness. Some amount of trial and error is usually required to a good pellet.

Common problems in making a KBr pellet

Cracked or partially formed pellet: Too small an amount of mixture in the press. Try again, using a larger amount.

Opaque (and often grainy) pellet: Too large an amount of mixture in the press. Try again, using a smaller amount.

Grainy pellet: Either too much mixture in the press or not enough pressure used to prepare the pellet. You should feel some resistance as the press is closed, but must not treat using the press as a muscle-building exercise! Forcing the press to close can damage it, and it is both surprisingly easy and expensive to damage a press in this way.

White or colored flecks in a transparent pellet: Compound was not sufficiently ground into the KBr. Try again, grinding more thoroughly before attempting to press the pellet.

Common problems in obtaining an IR spectrum

Strongly sloping baseline: Incomplete grinding. Try again, using more care in the initial grinding of the compound with KBr.

Noisy spectrum – high % transmittance values: Not enough compound. Try again, using more compound when preparing the pellet or using a thicker pellet. For instruments using automatic subtraction of a "background" spectrum, ensure that the background spectrum is appropriate.

Noisy spectrum – low % transmittance values: Too much compound or pellet too thick. Try again, using less compound when preparing the pellet or using a thinner pellet. (Simply regrinding the pellet with a little more KBr, then pressing again, can work well.) For instruments using automatic subtraction of a "background" spectrum, ensure that the background spectrum is appropriate.

Upside-down peaks: Ensure that the instrument is set to record % transmission, not absorbance. For instruments using automatic subtraction of a "background" spectrum, ensure that the background spectrum is appropriate.

All peaks broad and featureless: Wet sample. Try again, making sure your compound and the KBr are completely dry.

EXPERIMENT 7
GAS-PHASE SYNTHESIS, COLUMN CHROMATOGRAPHY AND VISIBLE SPECTROSCOPY OF 5,10,15,20-TETRAPHENYLPORPHYRIN

Chemical Concepts

Synthesis of porphyrins; electrophilic aromatic substitution; gas-phase reactions; column chromatography; UV-visible spectroscopy; thin-layer chromatography.

Green Lessons

Solvent-free reactions; avoidance of corrosive reagents; oxidation by air.

Estimated Lab Time

2 hours

Introduction

Porphyrins or hemes (iron-containing porphyrins) play essential roles in living organisms. Indeed, life as we know it is in many ways dependent on porphyrins and their derivatives. In animals, for example, they are responsible for the oxygen-carrying action of hemoglobin and serve as essential electron transport "relay stations" in metabolism. In plants, closely related molecules form the basis of chlorophyll, serving as part of the light harvesting "antennae" in photosynthesis. Understandably, chemists have long been interested in the structure and synthesis of these important, but often complex molecules. The Nobel Prize in Chemistry was awarded to Hans Fischer in 1930 in recognition of his work with heme compounds, and Richard Willstätter earlier received the same honor, in 1915, for his studies of chlorophyll. The extensive conjugation within the porphyrin ring leads to intense absorption in and near the visible light region, and as a result porphyrins are typically dark-red to purple in color.

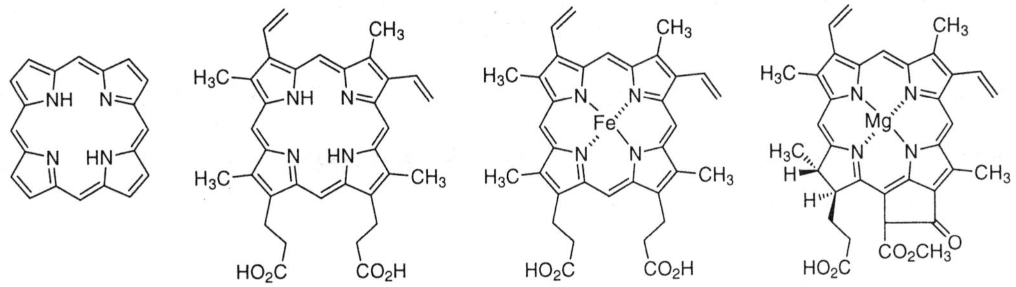

porphyrin protoporphyrin IX heme chlorophyllin

Typical porphyrin and porphyrin-like compounds

Nearly all syntheses of porphyrins exploit electrophilic aromatic substitution reactions, in which an electrophile replaces (substitutes for) a hydrogen substituent on an aromatic ring. This type of reaction, explored in greater detail for simple benzene derivatives in Experiment 12, is illustrated here with pyrrole serving as the aromatic compound undergoing electrophilic substitution.

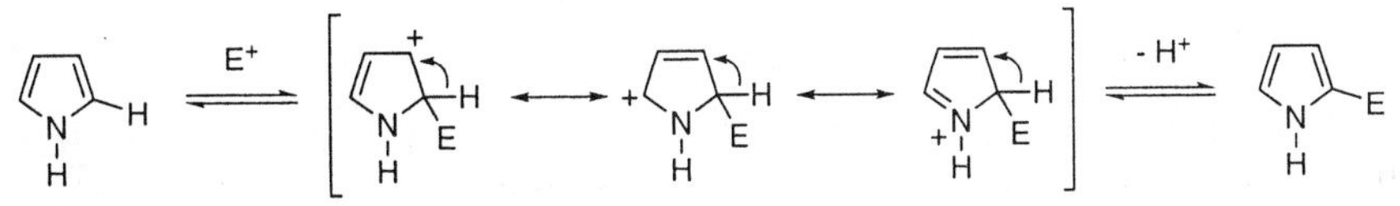

pyrrole (resonance-stabilized intermediate)

Pyrrole undergoes typical electrophilic aromatic substitution reactions

In this experiment, you will prepare a simple porphyrin, 5,10,15,20-tetraphenylporphyrin. The carbon framework of the porphyrin is assembled from benzaldehyde and pyrrole through *eight* electrophilic substitution reactions, resulting in the formation of eight new carbon-carbon bonds. The initial product is not a porphyrin, but rather a porphyrinogen, representing a reduced porphyrin, with saturated methylene groups separating the pyrrole fragments. Air oxidation of the porphyrinogen yields the ultimate porphyrin product, which is a fully-conjugated, 18-electron aromatic compound.

Mechanism of porphyrinogen and porphyrin formation from benzaldehyde and pyrrole

Porphyrin syntheses have traditionally been carried out in corrosive, high-boiling solvents such as propanoic acid. Alternative syntheses generally employ large volumes of halogenated solvents containing corrosive Lewis acid catalysts. Frequently, toxic reagents are used to oxidize the porphyrinogen to the porphyrin. In this experiment, you will employ a method of porphyrin synthesis, originally developed by Drain and Gong [49] and modified by Hutchison, *et al.* [50], that requires no solvent or catalyst. The two reagents, pyrrole and benzaldehyde, react at high temperature in the gas phase to form the porphyrinogen intermediate, which is then oxidized to the porphyrin product by oxygen (provided by air). This gas-phase experiment demonstrates a method for the eliminating hazardous solvents in the preparation of organic compounds and illustrates the avoidance of corrosive reagents and the use of air as a mild oxidant (c.f. Experiment 6 for additional discussions of this point). The isolation and purification procedures for this experiment are also greener than those for traditional syntheses, which generally require chlorinated solvents such as dichloromethane or chloroform. Here,

49.   C. M. Drain and X. Gong, "Synthesis of *meso* substituted porphyrins in air without solvents or catalysts," *Chem. Commun.* **1997**, *42*, 2117.
50.   M. G. Warner, G. L. Succaw, and J. E. Hutchison, "Solventless syntheses of meso-tetraphenylporphyrin: New experiments for a greener organic chemistry laboratory curriculum," *Green Chemistry* **2001**, 267-270.

the crude product simply deposits on the walls of the reaction vessel and may be purified by column chromatography using a safer solvent mixture (hexanes and ethyl acetate).

Pre-Lab Preparation

1. Read the article by Drain and Gong (*Chem. Commun.* **1997**, *42*, 2117) describing this synthetic approach.

2. Study the technique sections in your lab manual regarding column chromatography, TLC, visible spectroscopy, and use of the rotary evaporator.

3. Carry out pre-lab preparations as described in Chapter 11, section 11.6A, or as called for by your instructor.

Experimental Procedure

SAFETY PRECAUTIONS: Ethyl acetate, hexanes, and acetone are flammable; avoid exposure to open flames. Avoid inhalation of silica gel particles or fumes of benzaldehyde or pyrrole. Pyrrole and benzaldehyde can be irritating to the skin – avoid contact. The reaction is effected at a relatively high temperature – take care to avoid thermal burns.

Reaction

1. Cap a 5 mL conical vial with a septum-bearing screw cap. Place the vial in a heating mantle filled with sand so that the bottom 1.5 inches of the vial is submerged in the sand. Apply heat to the sand bath, and when the temperature of the bath reaches 170 °C, use a 20 μL syringe to inject 10 μL of benzaldehyde (0.1 mmol) through the septum. (Test the syringe before use to ensure that it is not blocked – try squirting a bit of benzaldehyde onto a piece of tissue. Be sure to clean the syringe after you use it.)

2. Once droplets of benzaldehyde form on the walls of the vial and the temperature has reached approximately 180 °C, use a 10 µL syringe to inject 7 µL of pyrrole [51] (0.1 mmol) through the septum.

3. Raise the temperature of the sand bath to 250°C. After heating the vial for 15 minutes at 250 °C, remove it from the sand bath and allow it to cool to room temperature on the bench top.

Workup

4. After the vial is cool, use 1 mL of CH_2Cl_2 to rinse the cap liner and walls of the vial, affording a solution of the crude product, ready for chromatographic analysis and purification.

Thin Layer Chromatography (TLC)

5. Analyze the product mixture by thin layer chromatography on a silica TLC plate, using a 7:1 hexanes/ethyl acetate mixture to elute the plate. (Remember to make sure all the spotting solvent has evaporated before developing the plate to avoid irreproducible results based on the presence of traces of CH_2Cl_2.) Tetraphenylporphyrin, easily recognizable by its violet color, should appear as the leading spot on the silica plate, with an R_f value of around 0.46. The remaining impurities appear as a broad band at lower R_f (0 to 0.3).

Purification – Column Chromatography

6. Prepare a silica gel column in a chromatography column (3 – 5 mm inside diameter) fitted with a Teflon stopcock. If the column contains a porous glass disk, add the silica directly, as described below. If not, insert a small plug of glass wool into the bottom of the column. Close the stopcock and add 5 mL of 7:1 hexanes/ethyl acetate, then gently add sand to a depth of about 2 cm. (The presence of the solvent aids in the formation of a uniform sand layer.) Prepare a loose slurry of approximately 6.5 g of silica gel in 7:1 hexanes/ethyl acetate (approximately 30 mL). Swirl the flask to make sure the silica gel is thoroughly suspended in the solvent, then quickly, but carefully,

---

51. Pyrrole should be passed through a short column of basic alumina (Al_2O_3) prior to use. If protected from light, the purified pyrrole may be stored for several days. Your instructor will provide purified pyrrole for you.

pour the suspension into the column. Open the stopcock and allow the silica gel to settle and the solvent to slowly drain until the level of the top of the silica gel stops lowering. Your column height should be around 30 cm. If it is not, add more slurry to the column. Finally, carefully deposit a 2 cm layer of sand to protect the top surface of the column. The sand should fall through the remaining solvent and make an even layer on the top of the silica gel.

7. Drain the solvent down to the top of the sand. Carefully load the entire solution of the product mixture in CH_2Cl_2 on the top of the column and elute until the solvent level has reached the top of the sand. Stop the flow and carefully add the elution solvent (7:1 hexanes/ethyl acetate). Run the column at a flow rate of approximately 30 drops/min until the leading purple band elutes. Generally, the entire porphyrin band can be collected in approximately 7 – 8 mL of solvent after a total of about 20 – 25 minutes.

Keys to successful column chromatography

- Carry out initial TLC to find the optimal elution solvent.
- Pack a good column – make sure every layer is uniform.
- Add the sample solution in as small a volume of solvent as possible.
- If performing gradient chromatography (you are not in this experiment), start with the least polar solvent and don't change the solvent composition too rapidly.
- If there is no obvious visual cue about the contents of eluted fractions (there is in this experiment – the desired compound is purple), collect fractions and analyze each one before mixing any of them together

---

Visible Spectroscopy

8. Prepare your sample for visible spectroscopy by placing 1 – 2 drops of the highly colored tetraphenylporphyrin solution in a sample vial and diluting it to 4 mL with additional 7:1 hexanes/ethyl acetate. Add a few drops of triethylamine to the solution to prevent protonation of the porphyrin by any traces of acid. **Note: In order to estimate your yield, you will need to know the volume of tetraphenylporphyrin solution used and the amount of solvent used to dilute the sample. The more carefully you measure and prepare your solutions, the higher the accuracy in your determination of yield.**

9. The UV/visible absorption spectrum of tetraphenylporphyrin shows a strong absorbance at 420 nm (extinction coefficient = $\varepsilon = 4.7 \times 10^5$) along with four weaker absorptions at 510 ($\varepsilon = 1.87 \times 10^4$), 550 ($\varepsilon = 7.7 \times 10^3$), 590 ($\varepsilon = 5.4 \times 10^3$), and 645 nm ($\varepsilon = 3.4 \times 10^3$). Adjust the concentration of your sample, keeping track of the volume of solvent added in order to permit an estimated yield calculation, until the absorbance at 420 nm is around one absorbance unit. Print a copy of this spectrum, labeling the wavelengths of the absorption maxima if possible. Also print a second copy, with the absorbance axis scaled to half of its original numerical maximum.

Post-Lab Questions and Exercises

1. To the best of your ability based on your spectroscopic analysis, using the extinction coefficients listed above, report the mass and percent of theoretical yield of the product.

2. Describe your TLC results for the reaction mixture. What were the R_f values for the spot(s)?

3. Describe what happened during column chromatography. What bands did you see elute, in what order?

4. Given an absorbance maximum at about 420 nm, what color would you *expect* the porphyrin to be? What color is the porphyrin that you isolated? How do you explain this color?

5. Attach your UV/visible spectra. Be sure to indicate which fraction from the column chromatography you used to obtain these spectra. Label the absorbance and wavelength of each peak in the spectrum.

6. Calculate the atom economy for the reaction.

Experiment Development Notes

This is another experiment taken from the recent primary literature [49]. Very few details were given in the original report, and many modifications were required in order to develop the method and optimize it for reliable use in the teaching laboratory [50]. This experiment is particularly appealing given the single-step formation of a complex organic molecule and the colorful column chromatography.

158

EXPERIMENT 8

MICROWAVE SYNTHESIS OF 5,10,15,20-TETRAPHENYLPORPHYRIN

Chemical Concepts

Electrophilic aromatic substitution; visible spectroscopy; column chromatography; thin-layer chromatography.

Green Lessons

Solvent-free reactions; solid-supported synthesis; microwave heating of reaction mixtures; safer solvents (for chromatography).

Estimated Lab Time

2.5 hours

Introduction

The preceding experiment introduced the chemistry of porphyrins and related compounds and discussed some of the chemistry involved in their synthesis. In this experiment you will explore an alternative solvent-free synthesis of 5,10,15,20-tetraphenylporphyrin from benzaldehyde and pyrrole. Instead of a gas-phase reaction at high temperature, you will use microwave irradiation to heat the reactants. In this case, the liquid reactants are adsorbed on a solid support, silica gel, which may act as a Lewis acid catalyst to facilitate the reaction. As in the preceding experiment, the reaction presumably forms a porphyrinogen, which is then oxidized to the porphyrin product by atmospheric oxygen.

Isolation and purification of the tetraphenylporphyrin product is effected by removal of the crude product from the silica, followed by column chromatography. As in the gas-phase synthesis procedure, this chromatography utilizes a safer solvent (a mixture of hexanes and ethyl acetate) than the halogenated solvents traditionally employed. This experiment, which clearly complements the preceding gas-phase synthesis, illustrates a number of other green chemical issues, including the avoidance of solvent usage, the use of solid-supported reactions, and the use of alternative energy sources (here, microwave energy) to effect chemical reactions.

Pre-Lab Preparation

1. Study the technique sections in your lab manual regarding column chromatography, TLC, visible spectroscopy, and use of the rotary evaporator.

2. Carry out pre-lab preparations as described in Chapter 11, section 11.6A, or as called for by your instructor.

Experimental Procedure

SAFETY PRECAUTIONS: Ethyl acetate, hexanes, and acetone are flammable; avoid exposure to open flames. Avoid inhalation of silica gel particles or fumes of benzaldehyde or pyrrole. Pyrrole and benzaldehyde can be irritating to the skin – avoid contact. When the reaction vessel is removed from the microwave oven, it will be very hot – take care to avoid thermal burns.

Reaction

1. Mix 0.43 mL of benzaldehyde and 0.3 mL of pyrrole in a 25 mL Erlenmeyer flask. Once the reactants are thoroughly mixed, add 0.63 g of silica gel, stopper the flask, and mix well until the silica gel is evenly and completely covered with the reactant mixture.

2. Place the flask containing the reaction mixture in the microwave oven (a standard 1000 W model), cover with a Pyrex watch glass, and heat for a total of 10 minutes in five 2-minute intervals.

3. Once the reaction is complete, allow the mixture to cool to room temperature, then add approximately 15 mL of ethyl acetate. Filter the solution to remove the silica gel, then remove the ethyl acetate using a rotary evaporator. Extract the residue with 1 mL of CH_2Cl_2 to prepare for chromatography.

Isolation and Characterization

4. Carry out thin layer chromatographic analysis of your crude reaction mixture, column chromatography to separate the tetraphenylporphyrin, and UV/visible spectroscopic analysis to estimate the yield of your product as described in the preceding experiment. If you plan to carry out the following experiment in this text – the metallation of 5,10,15,20-tetraphenylporphyrin – save the first three drops of porphyrin-containing solution eluted from your column chromatography.

5. This procedure provides tetraphenylporphyrin of approximately 85% purity, as determined by ^{1}H NMR integration. The major impurity can be seen in the ^{1}H NMR spectrum as a broad multiplet near 7.3 ppm. The impurity does not affect the UV/visible spectrum of the porphyrin, and the product is of satisfactory purity for use in the following experiment.

Post-Lab Questions and Exercises

1. To the best of your ability based on your spectroscopic analysis, report the mass and percent of theoretical yield of the product.

2. Describe your TLC results for the reaction mixture. What were the R_f values for the spot(s)?

3. Describe what happened during column chromatography. What bands did you see elute, in what order?

4. Attach your UV/visible spectra. Be sure to indicate which fraction from the column chromatography you used to obtain these spectra. Label the absorbance and wavelength of each peak in the spectrum.

5. Compare the "greenness" of this procedure with a more conventional synthesis in hot propanoic acid.

6. Assuming that this procedure affords a 5% yield, how much pyrrole, benzaldehyde and silica gel would you need to prepare 100 mg of tetraphenylporphyrin?

7. Most organic compounds display proton NMR resonances in the range of 0 – 10 ppm. In addition to a number of peaks in this "normal" region, the proton NMR spectrum of tetraphenylporphyrin

displays a resonance at –2.7 ppm (i.e., 2.7 ppm *upfield* from tetramethylsilane). Suggest which protons are responsible for this unusual resonance and explain why it appears so far upfield.

Experiment Development Notes

This experiment was developed, with extensive modification for use in the organic teaching lab [50], based upon the original report from Petit, *et al.* [52]. Although this preparation offers a number of advantages over traditional methods, there is room for further improvements, some of which could form the basis of student inquiry-driven investigation. Development of alternative chromatographic procedures that could reduce solvent usage during chromatography is desirable. (Although one of the primary goals of this experiment is to provide experience with column chromatography, eliminating the chromatography step is a possible option if this experience is not deemed necessary.)

Use of more benign benzaldehyde derivatives in place of benzaldehyde itself would also be desirable. This actually suggests an interesting possible extension of the experiment to the use of *ortho*-substituted benzaldehyde derivatives. The resulting *ortho*-substituted tetraphenylporphyrins will display interesting TLC and ^{1}H NMR spectral behavior due to restricted rotation (atropisomerism) [53]. This provides a platform for the discussion of more advanced topics in spectroscopy, such as the observation of temperature dependent phenomena and measurement of rates of interconversion by line broadening methods. Finally, the porphyrin product can also be used to construct a functioning solar cells according to a procedure published in the recent literature [54]. We hope to include this experiment in the next edition of this text.

52.   A. Petit, A. Loupy, P. Maillard, and M. Momenteau *Synth. Commun.* **1992**, *22*, 1137.
53.   R. F. Beeston, S. E. Stitzel, and M. A. Rhea, *J. Chem. Ed.* **1997**, *74*, 1468.
54.   E. N. Durantini and L. Otero, *Chem. Educator* **1999**, *4*, 144.

EXPERIMENT 9

METALLATION OF 5,10,15,20-TETRAPHENYLPORPHYRIN

Chemical Concepts

Coordination chemistry; visible spectroscopy.

Green Lessons

Safer solvents; elimination of need for heating.

Estimated Lab Time

4 hours

Introduction

It has long been known that metal salts react with neutral molecules such as water or ammonia to form compounds. The studies of Alfred Werner, leading to his receipt of the Nobel Prize in Chemistry in 1913, revealed that these compounds have well-defined structures, arising from the donation of at least one pair of electrons from the neutral molecule to the metal center. Organic molecules that can donate one or more electron pairs can also form such *"coordination complexes"* with metal ions. The molecule that donates the electron pair is called a *ligand*, and this definition has been extended to charged fragments (ions) as well. Common ligands include halides (e.g., Cl^-, Br^-), alkoxides (RO^-), amines (R_3N), and amides (deprotonated amines, R_2N^-). The nature of the ligands usually has a significant impact on the solubility and reactivity of the complex. In this experiment, for example, a simple water-soluble zinc salt is converted into a complex that is insoluble in water, but freely soluble in common organic solvents. Other properties of the metal ion – including redox behavior, Lewis acidity, and color – are also greatly influenced by the nature of the attendant ligands.

Coordination complexes are important throughout chemistry and biology. They act as catalysts, drugs, oxygen carriers, and electron transfer agents, in addition to playing myriad other roles. Metal complexes of the porphyrin ring system, introduced in the preceding experiments, are central to numerous biochemical functions. The iron-containing hemes are ubiquitous in biological systems, acting as essential components in the electron transport chain and serving as the oxygen binding site in myoglobin (the oxygen storage protein) and in hemoglobin (the oxygen transporting protein), while a magnesium porphyrin complex forms the basis of chlorophyll.

In addition to the "metalloporphyrins" (i.e., metal complexes of porphyrins) found in biological systems, many synthetic metal complexes of porphyrins have been prepared by chemists for use as models of the more complex biological systems and for use as catalysts, sensors, and other functional materials. Usually, the metal-free porphyrin ligand is synthesized and purified prior to insertion of the metal (metallation). Metallation of the porphyrin ring requires removal of two protons, and in some cases the metallation is reversible, with removal of the metal effected by heating with acid. Virtually every metal in the periodic table has been inserted into a porphyrin; some form complexes very easily, while others require more forcing conditions. In this experiment, you will use zinc(II) acetate $[(CH_3CO_2)_2Zn]$ to metallate tetraphenylporphyrin, monitoring the metallation using visible spectroscopy.

Metallation of 5,10,15,20-tetraphenylporphyrin with zinc(II) acetate

Typically, halogenated solvents or N,N-dimethylformamide (DMF) are used as solvents in porphyrin metallation reactions. In this experiment, you will avoid the use of these solvents, using instead two more benign solvents, 1-methyl-2-pyrrolidinone (N-methylpyrrolidinone, NMP) and dimethylsulfoxide (DMSO). Whereas porphyrin metallation often requires elevated temperatures, this solvent mixture allows the reaction to proceed at a reasonable rate at room temperature.

Pre-Lab Preparation

1. Study the technique section in your lab manual regarding visible spectroscopy.

2. Carry out pre-lab preparations as described in Chapter 11, section 11.6A, or as called for by your instructor.

Experimental Procedure

> SAFETY PRECAUTIONS: DMSO can readily penetrate the skin and may bring along any compounds dissolved in it or any contaminants that might be on your skin. Wear gloves and avoid getting DMSO solutions on your skin.

Note: If this experiment is being performed in combination with either the gas-phase synthesis (Experiment 7) or the microwave synthesis of tetraphenylporphyrin (Experiment 8), collect three drops of the leading (porphyrin) fraction from the column chromatography, evaporate to dryness, and dilute to 4 mL with N-methylpyrrolidinone (NMP). If using commercially obtained tetraphenylporphyrin, dissolve a trace (no more than 1-2 mg) in 4 mL of NMP.

Reaction

1. Record an initial visible spectrum of the tetraphenylporphyrin solution in a clean, dry glass cuvette. There is a very strong absorbance at 420 nm and four weaker absorbances (the so-called Q-bands) at 510, 550, 590, and 645 nm. As needed, adjust the concentration of your porphyrin solution to allow you to observe the weaker Q-bands. It is the weaker 3rd and 4th Q-bands, which are not present for the zinc complex, that are most diagnostic of the metallation reaction.

2. Dissolve 0.400 g of $(CH_3CO_2)_2Zn$ (2.2 mmol) in a minimum amount of DMSO. **Note: Your instructor may choose to prepare this solution for you. If so, proceed directly to step 3.**

3. Add 5 drops of the $(CH_3CO_2)_2Zn$ solution to the porphyrin solution contained in the glass cuvette and mix thoroughly. Record a spectrum every twenty-five minutes to monitor the progress of the metallation reaction. The metallation is typically complete in approximately four hours at room temperature.

Post-Lab Questions and Exercises

1. Attach your UV/visible spectra. Label the absorbance and wavelength of each peak.

2. Report any color changes that you observed during the course of the reaction. Explain how these color changes correlate with the spectral changes that you observed.

Experiment Development Notes

This experiment represents a modification [50] of the procedure reported in the primary literature by Petit, *et al.* [55]. Replacement of DMSO may be desirable; the challenge is to find a solvent or solvent mixture in which both tetraphenylporphyrin and the zinc salt are soluble. With the instructor's approval, students may wish to explore the formation of complexes of tetraphenylporphyrin with other metals. In such a study, be aware that some complexes form appreciably less rapidly than the zinc complex.

55. A. Petit, A. Loupy, P. Maillard, and M. Momenteau *Synth. Commun.* **1992**, *22*, 1137.

EXPERIMENT 10

MEASURING SOLVENT EFFECTS:
KINETICS OF HYDROLYSIS OF *TERT*-BUTYL CHLORIDE

Chemical Concepts

S_N1 reactions; solvent effects; chemical kinetics; titration.

Green Lessons

Alternative solvent selection – recognition of impact of solvent choice on reaction rate.

Estimated Lab Time

1.5 – 2 hours

Introduction

The solvent in which a reaction is carried out often influences the rate of the reaction. If the reactant is more stabilized by interaction with solvent molecules than is the transition state, the activation barrier for the reaction will be increased, and the rate will decrease. Conversely, if the transition state is more stabilized by solvation than is the reactant, the activation barrier for the reaction will be decreased, and the rate will increase. Although solvents can interact with reactants (and transition states) in a number of ways, two particularly significant interactions are the solvation of ions by polar solvents and hydrogen bonding between anions and hydrogen bond donating solvents.

In this experiment, you will explore these issues by measuring the kinetics of a unimolecular substitution reaction – the S_N1 reaction between water and *tert*-butyl chloride (2-chloro-2-methylpropane) in various solvents. The energy diagram depicting the course of this hydrolysis reaction, provided below, serves as a convenient reference for consideration of solvent effects on the rate of this reaction.

167

$$R_3C-X \quad \underset{\text{slow}}{\rightleftharpoons} \quad R_3C^+ + X^- \quad \underset{H_2O}{\rightleftharpoons} \quad R_3C-OH_2^+ + X^- \quad \underset{-HX}{\rightleftharpoons} \quad R_3C-OH$$

Hydrolysis of tert-butyl chloride – an S_N1 substitution reaction

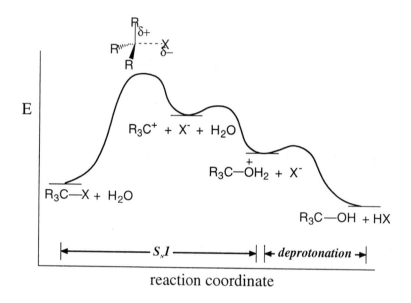

reaction coordinate

Energy diagram for the hydrolysis of tert-butyl chloride

In the first transition state, associated with the slower (rate-determining) step, the carbon-chlorine bond is partially broken; the chlorine bears a partial negative charge and the carbon a partial positive charge. The charge-separated nature of the transition state will be better stabilized by polar solvents than by nonpolar solvents. Thus, ignoring any differences in solvation of the reactant, increasing the polarity of the solvent – better solvating the polarized transition state and thereby lowering the activation barrier – should increase the rate of this hydrolysis reaction. Similarly, moving to a better hydrogen bond-donating solvent should allow stabilization of this transition state through hydrogen bonding with the chlorine, bearing a partial negative charge; again, the rate of the hydrolysis reaction should be enhanced.

168

As has been discussed (Chapter 6), a common strategy in reducing or eliminating hazards in a synthesis involves replacing a hazardous solvent with a more benign one. In making such substitutions, it is important to know that changing the solvent can lead to dramatic changes in reactivity – switching to a greener solvent may either hinder or enhance the desired reactivity. Similar issues arise when contemplating selection of an alternative reactant (Chapter 9) or reagent (Chapter 7). This experiment is designed to emphasize these points.

Pre-Lab Preparation

1. Carry out pre-lab preparations as described in Chapter 11, section 11.6A, or as called for by your instructor.

Experimental Procedure

SAFETY PRECAUTIONS: Acetone, 2-propanol, and *tert*-butyl chloride are flammable liquids. Avoid open flames.

In this experiment you will be studying the kinetics of hydrolysis of *tert*-butyl chloride in one of four different solvent mixtures – 50/50 acetone/water, 60/40 acetone/water, 50/50 2-propanol/water, and 60/40 2-propanol/water. Your instructor will assign a solvent mixture for you. By sharing your results with those of your classmates using other solvent mixtures and considering the class's collective data set, you will be able to analyze the effect of solvent composition on the rate of this reaction.

Note: The accuracy of your results depends on the experimental care you take. Pay particular attention to accurate weighing of materials, measuring volumes, monitoring reaction times, performing titrations, and keeping track of units in your calculations. Remember, your classmates are relying on you to work carefully and accurately.

Reaction

1. Obtain a stopwatch, a dropping bottle of bromothymol blue indicator solution, and a white card (to provide a plain white background when determining endpoints by titration).

2. Prepare 100 mL of the appropriate solvent mixture (assigned by your instructor) in an Erlenmeyer flask. Make sure the solvents are completely mixed, then stopper the flask and set it aside, allowing it to come to ambient laboratory temperature. Record the laboratory temperature in your notebook.

3. In a 250 mL Erlenmeyer flask, place 150 mL of 0.04 N standardized aqueous NaOH solution. Using this solution, fill your burette.

4. Weigh 1 g (± 0.01g) of *tert*-butyl chloride into a 100 mL volumetric flask, then dilute to the line with your solvent mixture. Stopper the flask, shake well, and start your stopwatch. Except when removing aliquots of this reaction mixture, keep this flask stoppered to prevent evaporation.

5. Prepare two "infinity" time samples by adding a 10 mL sample of your reaction mixture to each of two Erlenmeyer flasks containing 10 mL of water. Stopper these flasks securely and let them stand at least one hour while you complete the next steps. The high concentration of water in these samples drives the reaction to completion relatively rapidly, so that they can be used to determine the concentration of HCl present when the reaction is complete ("infinite time"), which will be represented as [HCl]$_\infty$. By working with two samples, you decrease the possibility of being confused by experimental error.

6. Remove aliquots of the reaction mixture at the time intervals indicated below for your solvent. At each sample time, withdraw a 10 mL sample and place it in a 125 mL Erlenmeyer flask containing 15 mL of acetone. The hydrolysis reaction is quite slow in acetone, so this effectively "quenches" the reaction. Record the exact time (t) that you quench each sample. Add 3 drops of bromothymol blue indicator solution to the flask and titrate with the standardized NaOH solution to a blue end point that persists for 20 seconds. Record the volume of NaOH required to reach this end point.

Be sure to clean and dry the sampling pipette after withdrawing each aliquot. Data for the 50/50 solvent mixtures should be collected every 15 minutes up to 90 minutes, and then a final data point should be collected at 115 minutes. For the 60/40 mixtures, collect data every 20 minutes up to 120 minutes.

7. After your data collection is complete, titrate the two infinity time samples in the same way. Record the volume of NaOH solution required to reach the end point.

Data analysis

8. From your titrations of the "infinity" time samples, calculate $[HCl]_\infty$.

9. From your titrations of the aliquots of the reaction mixture, calculate the concentration of HCl present at each of the sampling times (t), which will be represented as $[HCl]_t$.

10. Plot $\ln([HCl]_\infty - [HCl]_t)$ vs. time (t). (Refer to the mathematical background notes following this experiment if the reasons for constructing this plot are not clear to you.) Draw the best straight line through the points and calculate the slope. From the slope, calculate the rate of the reaction. Report your rate and solvent composition to your instructor, and obtain rate data obtained by your classmates, both for your own and for the other solvent mixtures.

Post-Lab Questions and Exercises

1. Provide the plot of your data as described above and a table summarizing your kinetic data.

2. Compare the rate that you obtained to that reported by other students in your section by preparing a table comparing rates and solvent mixtures.

3. How does your determined rate compare with that by your classmates for the same solvent compositions?

4. How do the rates found at different solvent compositions compare?

5. Is the solvent dependence consistent with the proposed mechanism for the S_N1 reaction?

6. In the solvolysis of *tert*-butyl chloride, some *tert*-butyl ether is formed. Do you think this will affect the kinetics of the reaction?

7. Prepare energy diagrams for the hydrolysis of *tert*-butyl chloride (S_N1 reaction) with two different concentrations of water. Explain any significant differences in the diagrams and how they affect the reaction rate.

Experiment Development Notes

This represents a classic kinetics experiment that has appeared in some form or another in many organic laboratory texts. Here, we have expanded the range of solvents used and have configured the experiment as a cooperative investigation, where students each investigate a different solvent, share their results with the rest of the class, and reach conclusions based upon the cumulative data of the class.

Mathematical Background

The reaction being studied can be summarized as follows, where k is the rate constant for the forward reaction.

$$R-Cl \underset{k}{\overset{H_2O}{\rightleftharpoons}} R-OH + HCl$$

At time t = 0, $[RCl] = [RCl]_0$ and $[HCl]_0 = 0$, since no R-Cl has yet reacted.

At time t > 0, $[RCl]_t = [RCl]_0 - [HCl]_t$.

At time t = ∞, all of the RCl has reacted to form HCl, so that $[RCl]_0 = [HCl]_\infty$. Since it is easier and more accurate for us to measure $[HCl]_\infty$ than $[RCl]_0$, we use equation (1).

$$[RCl]_t = [HCl]_\infty - [HCl]_t \tag{1}$$

As equation (2) states mathematically, the rate of formation of ROH is equal to the rate of loss of RCl.

$$\frac{d[ROH]_t}{dt} = -\frac{d[RCl]_t}{dt} \tag{2}$$

These rates are dependent on the concentrations of ROH and RCl. This is inconvenient because the concentrations will change as the reaction progresses, leading to a non-linear rate versus t correlation.

172

To compare the rates of reactions, we need a constant that is independent of concentration – this is called the rate constant (k) and used as indicated in equation (3).

$$-\frac{d[RCl]_t}{dt} = k \times [RCl]_t \qquad (3)$$

Equation (3) may be rewritten into a standard form – equation (4) - that can be integrated.

$$\frac{d[RCl]_t}{[RCl]_t} = -k\, dt \qquad (4)$$

Integrating from 0 to t on both sides [equation (5)] affords equation (6).

$$\int_{[RCl]_0}^{[RCl]_t} \frac{d[RCl]_t}{[RCl]_t} = \int_0^t k\, dt \qquad (5)$$

$$\ln \frac{[RCl]_t}{[RCl]_0} = -kt \qquad \underline{(6)}$$

Finally, substituting equation (1) into equation (6) results in equation (7), which is of the linear form $y = mx + c$. Thus, plotting $\ln([HCl]_\infty - [HCl]_t)$ against t should yield a straight line with a y-intercept of $\ln[HCl]_\infty$ and a slope of $-k$ (in units of s^{-1}).

$$\ln([HCl]_\infty - [HCl]_t) = \ln[HCl]_\infty - kt \qquad (7)$$

EXPERIMENT 11
MOLECULAR MECHANICS MODELING

Chemical Concepts

Computational methods; molecular mechanics calculations; molecular modeling.

Green Lessons

Design of safer products; synthetic efficiency.

Estimated Lab Time

2 – 3 hours

Introduction

Organic chemists frequently synthesize a large number of compounds in order to "screen" them for desired properties. This is particularly true in the pharmaceutical industry, where discovery of an initial "lead" compound displaying promising activity (e.g., antibacterial or anticancer activity) can lead to the synthesis of thousands of analogs in an attempt to improve efficacy and reduce toxicity or undesirable side effects. As our understanding of molecular structure and function has grown, and as more and more powerful computational resources (e.g., national supercomputer centers) have become available, it has become increasingly possible to use computational methods to explore and predict the properties of unknown molecules in advance of their synthesis. Such computational methods are often referred to as "molecular modeling."

Through the years, various computational approaches have been developed for molecular modeling. "Molecular mechanics" calculations are particularly useful in many cases, and are relatively easy to understand conceptually, making them an attractive place to begin one's investigations of computational chemistry. Molecular mechanics is a computational method of modeling molecular geometry and conformation in which the molecule is treated as a collection of balls (atoms) and springs (bonds). Each interaction in the collection of balls and springs is described by a mathematical equation. The total steric or strain energy of any given conformation of a molecule is calculated as the sum of the energies of all of the bond stretches and bends, torsional interactions, and non-covalent effects such as van der Waals and dipolar interactions. The conformation is then changed and the total energy recalculated, and this process is continued until the lowest energy (i.e., the most stable) conformation is determined. While these calculations are relatively simple to carry out, there are several factors that must be kept in mind.

- Molecular mechanics calculations yield only steric or strain energies. They do *not* take into account the *electronic* structure of the molecule unless parameterized to do so.

- It is possible to mistakenly identify a "local" minimum rather than a "global" minimum energy conformation. Care must be taken to ensure that one has really located the lowest energy conformation.

- These calculations do not take into account the effects of solvent and are thus most valid for molecules in the gas phase.

- While in many cases the computational results come very close to experimental values, these calculations represent approximations and thus should not be expected to predict exactly the structures and energies of molecules.

Computational approaches represent valuable additions to the "arsenal" of tools for green chemistry, in that they enable the exploration of many properties of molecules without the need to synthesize them. This can help to avoid the needless synthesis and screening of numerous compounds that turn out to be "duds," thereby reducing both waste generation and potential exposure to compounds of unknown biological activity. It is important to note, however, that there are limitations to the predictive capabilities of current computational methods. While these methods represent useful planning tools, helping to guide the synthetic chemist to specific target molecules holding the greatest promise, they are not yet capable of fully replacing synthesis and screening approaches. (Experiment 19 introduces a modern approach to the rapid synthesis and screening of many compounds known as combinatorial chemistry.)

In this experiment, you will use a commercially available molecular modeling software package, SPARTAN, to investigate the structures of several of the molecules featured in the preceding laboratory experiments.

Pre-Lab Preparation

1. Study the following overview of molecular mechanics calculations and the experimental section.

2. Carry out pre-lab preparations as described in Chapter 11, section 11.6A, or as called for by your instructor. Your preparations will obviously be somewhat different than for laboratories involving wet chemical procedures, but make sure that, when you sit down at the computer, you are ready to begin work.

A Brief Overview of Molecular Mechanics

Molecular mechanics calculations are commonly used to predict equilibrium molecular geometries or conformations. Mathematical expressions are used to represent various components of the bonding interactions (bond stretching, bond bending, and torsional interactions) and non-bonding interactions (van der Waals and electrostatic) that influence molecular geometry. Using these expressions, the total energy of a molecule, in a given conformation, is calculated. The simplest interactions to model are the bond stretching and bending energies. In a molecular mechanics calculation, atoms are treated as hard spheres interconnected by springs that represent the bonds, and thus stretching and bending can be quantified mathematically by Hooke's Law [equations (1) and (2)]. Energies calculated in this way will be lowest when bond lengths and angles are close to normal values and will increase as bonds are distorted from their normal equilibrium values (i.e., as strain is introduced).

$$E_{stretch}(r) = \frac{1}{2} k_{stretch}(r - r_{eq})^2$$

$$(1)$$

r = bond length, r_{eq} = ideal (equilibrium) bond length, $k_{stretch}$ = stretching force constant

$$E_{bend}(\alpha) = \frac{1}{2} k_{bend}(\alpha - \alpha_{eq})^2 \qquad (2)$$

α = bond angle, α_{eq} = ideal, equilibrium bond angle, k_{bend} = bending force constant

Analogous expressions may be formulated to approximate the torsional and non-bonded interactions as well, and the total energy of the molecule is then given by equation (3).

$$E_{total} = \overset{bonds}{\sum E_{stretch}} + \overset{bond\ angles}{\sum E_{bend}} + \overset{torsion\ angles}{\sum E_{torsion}} + \overset{noncovalent\ interactions}{\sum E_{noncovalent}} \qquad (3)$$

The conformation of the molecule that leads to the lowest overall energy is taken to be the equilibrium geometry. Development and testing of mathematical expressions that result in models best describing the experimentally measured properties of actual molecules represents an ongoing area of research.

Molecular mechanics calculations involve an iterative process of energy calculation, perturbation of the molecular conformation, and recalculation of total energy, continuing until a minimum energy conformation is obtained. This process may be outlined as follows.

1. Enter the atomic coordinates of the starting structure into the program. This is often achieved through a graphical interface, enabling input by simply providing the structure of the molecule. In this step, the atom types (carbon, hydrogen, etc.) are set and initial values for bond lengths, bond angles, and dihedral angles are established.
2. Calculate the energy of the structure.
3. Change the structure (e.g., by altering a bond length or angle) and recalculate the energy. If the energy is lower, the model has taken a step in the right direction, toward a minimum energy (equilibrium) structure.
4. Repeat step #3 until the lowest energy conformation is found. Such a structure is said to be *minimized*. Each round of structural change/calculation is called an *iteration*.

This procedure will result in a minimized structure. This *may* represent the lowest energy structure possible, a so-called *global minimum*. However, it is also possible that the structure is merely a *local minimum*, lower in energy than any of its closely related conformations, but higher in energy than the global minimum. The energy diagram for the rotation of the carbon-carbon bond in 1-chloro-2-fluoroethane provides a simple illustration of this point – the global minimum occurs at a dihedral angle of 180°, while local minima are present at rotations of 60° and 300°.

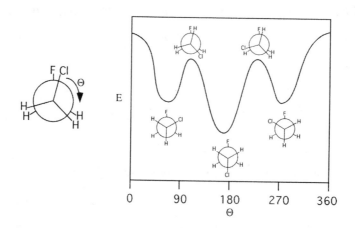

Energy diagram for rotation of carbon-carbon bond in 1-chloro-2-fluoroethane,
displaying global minimum at 180° and local minima at 60° and 300°

Thus, molecular mechanics calculations must be used with care. Two safeguards can improve the chances of obtaining a global minimum. First, restart the calculation from a drastically different conformation to see if the same minimized energy structure is obtained. If it is, there is a good chance it represents the global minimum. Second, use common sense, checking the structure to see if the bond lengths and angles appear reasonable. If they do not, this is a good indication that the structure represents only a local minimum. Finally, remember that the results of the computations are only models and these models are only as good as the method you used to calculate them. It is best to test any modeling program by using it to calculate the structure (bond lengths and angles) of a known compound before using it in an attempt to predict the unknown properties of a new compound.

Experimental Procedure

There are a variety of molecular modeling programs available for use on personal computers. The following experimental description is for Mac SPARTAN Plus v. 1.2.2 (Wavefunction, Inc., Irvine,

California, USA). Minor modifications are required when using PC SPARTAN Plus v. 2.0. If you are using a different version or different modeling program, refer to the instruction manual and/or receive information from your instructor regarding its usage.

CIS- AND TRANS-STILBENE

1. Build and minimize a model of *trans*-stilbene by carrying out the following steps.

   Choose **New** from the File menu.

   Using the Builder menu, construct *trans*-stilbene. Important hint: If you start with a completely planar structure, the program will find only a local minimum. Begin with a nonplanar structure in order to facilitate location of the global minimum.

   Minimize the structure in the Builder Menu (**Minimize** button); click **Done** when complete.

   Choose **Calculation** from the **Setup** menu.

   In the dialog box, choose Task = Geometry Optimization, Level = Sybyl, Charge = 0, Multiplicity = 1 and leave Options blank.

   Choose **Submit** in the Setup menu.

   Wait until the energy minimization is complete.

   Explore the different display modes in the *Model* menu.

   Record the energy (found in the *Output* submenu of *Display*) of the minimized structure for *trans*-stilbene. (The energy should be less than 7.)

   Under *File* menu, choose *Close*.

2. Repeat the above process for *cis*-stilbene. Again, the energy of your minimized structure should be less than 7.

ISOMERS OF 4-METHYLCYCLOHEXENE

3. Following the same steps, build and minimize both isomers of 4-methylcyclohexene. Make sure that you have found the lowest energy conformation for each isomer. Record the final energies of the two isomers.

1,4-BIS(1-HYDROXYCYCLOHEXYL)-1,3-BUTADIYNE

4. Build and minimize 1,4-bis(1-hydroxycyclohexyl)-1,3-butadiyne. Record the final energy.

5. By exploring the minimized structure, locate the longest atom-to-atom distance in the molecule. Record the identity of the atoms and the distance.

5,10,15,20-TETRAPHENYLPORPHYRIN DIANION

6. Build and minimize 5,10,15,20-tetraphenylporphyrin dianion (no hydrogens on any of the central nitrogen atoms, leading to a charge = 2-). In order to ensure that all of the ring atoms are sp² hybridized, use the phenanthryl ring system as your basic building block, then edit appropriately. This will take some effort; consult with your instructor if you have problems. (Some programs may provide a pre-formed porphyrin ring structure. If so, consider yourself lucky and continue with the exercise, adding the four phenyl rings to the structure.) Record the final energy.

7. Examine the minimized structure to determine if the porphyrin ring (ignoring the four phenyl substituents) is planar. (It should be.)

8. Measure and record the torsional angle between the porphyrin plane and each of the attached phenyl rings.

Post-Lab Questions and Exercises

1. What were your calculated energies for *cis-* and *trans-*stilbene? Do the relative energies of these two isomers make sense?

2. What were your calculated energies for the isomers of 4-methylcyclohexene? Do the relative energies of these two isomers make sense?

3. What is the longest atom-to-atom distance in 1,4-bis(1-hydroxycyclohexyl)-1,3-butadiyne? Which atoms are separated by this distance?

4. What was your calculated energy for 5,10,15,20-tetraphenylporphyrin? Was your minimized porphyrin ring system planar?

5. What were your measured torsional angles between the porphyrin plane and each of the attached phenyl rings in your minimized structure of 5,10,15,20-tetraphenylporphyrin? Why do you think the phenyl rings adopt this orientation?

Experiment Development Notes

This module was developed at the University of Oregon. It is easily adaptable to other computational packages, and clearly lends itself to student exploration of other molecules and other capabilities of the program. If your computational package includes other capabilities and time allows, a number of other interesting explorations can be carried out. For example, one can use AM1, a semi-empirical electronic structure calculational method, to simulate infrared spectra.

EXPERIMENT 12

ELECTROPHILIC AROMATIC IODINATION

Chemical Concepts

Electrophilic aromatic substitution; recrystallization; melting point determination.

Green Lessons

Safer and easier to handle reagents and solvents, more selective reagents.

Estimated Lab Time

3 hours

Introduction

In contrast to the substitution chemistry of alkyl compounds, which is dominated by nucleophilic substitution processes, aromatic compounds most often undergo *electrophilic* substitution reactions. In general, these reactions occur via a two-step addition/elimination mechanism, as introduced in Experiment 7. The electrophilic reagent first adds to the aromatic ring by attacking the π electrons, forming a cationic intermediate. This intermediate then eliminates a leaving group, often simply a proton, to form the substituted product.

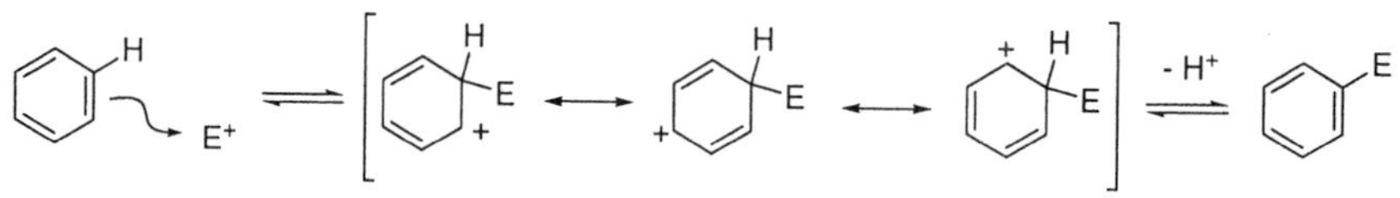

Two-step (addition/elimination) mechanism for aromatic substitution by an electrophile (E^+)

Consideration of this mechanism allows one to predict the effects of pre-existing aromatic ring substituents on the reactivity and regiochemistry of the substitution reaction. Thus, for example, electron-releasing substituents accelerate the reaction and preferentially drive the incoming substituent into the *ortho* and *para* positions, where resonance stabilization of the positive charge is optimized.

182

Halogenation of aromatic compounds (i.e., the replacement of a hydrogen atom by one of the group VIIA elements) is typically carried out through the reaction of the aromatic compound with Cl_2, Br_2, or I_2, with a catalyst typically required to mediate the reaction. Although elemental iodine is relatively easy to work with, bromine is a volatile liquid, and both the liquid and the vapors can cause serious chemical burns to the skin and eyes as well as severe to fatal irritation of the respiratory passages. Similarly, chlorine is a gas (bp -34 °C) that can cause severe to fatal respiratory irritation. Suffice it to say that we prefer not to use bromine or chlorine in the undergraduate laboratories. None-the-less, aromatic halogenation is an important tool for the elaboration of aromatic compounds into specific compounds of interest or importance.

Given the relatively low reactivity of elemental iodine toward many aromatic systems, a mixture of iodine and a powerful oxidizing agent (e.g., nitric acid) is often used to effect aromatic iodination reactions. If the aromatic substrate bears strongly electron-donating groups (e.g., phenols), the iodination reaction is more facile, but it can be difficult to control the regiochemistry of the reaction and polyiodination is common. In this experiment, you will explore the use of a convenient alternative iodination procedure, using sodium or potassium iodide instead of iodine and common bleach (sodium hypochlorite, NaOCl) as an oxidizing agent in aqueous alcohol. These reaction conditions offer two advantages over more traditional methods. The reaction is efficient and selective, leading to good yields of monoiodinated product, and the method allows the use of more environmentally-benign reagents (NaOCl instead of HNO_3) and solvents (aqueous alcohols instead of halogenated solvents).

Two procedures are presented in this experiment, one for the iodination of 4-hydroxyacetophenone, the other for vanillin (4-hydroxy-3-methoxybenzaldehyde). The latter procedure is more forgiving with regard to success even in the event of minor experimental errors than the former. The product from either reaction may be used in Experiment 13.

4-hydroxyacetophenone

Electrophilic iodination of 4-hydroxyacetophenone

Electrophilic iodination of vanillin (4-hydroxy-3-methoxybenzaldehyde)

Pre-Lab Preparation

1. Study the technique section in your lab manual regarding rotary evaporators, recrystallization, and melting points.

2. Carry out pre-lab preparations as described in Chapter 11, section 11.6A, or as called for by your instructor.

3. In the reaction workup, why does the addition of hydrochloric acid cause the product to precipitate?

4. Using structures to illustrate your answer, explain why iodination occurs preferentially *ortho* to the hydroxyl group rather than *ortho* to the acetyl or formyl group.

5. When carrying out the recrystallization of a crude solid reaction product, a cautious chemist will always retain a small amount of the crude solid, rather than dissolving it all. Why? (HINT: It is *not* to ensure that all the product is not lost if recrystallization is unsuccessful, since recrystallization is not a destructive procedure – i.e., one can always simply evaporate the solvent to recover the product if recrystallization fails.)

Experimental Procedure

SAFETY PRECAUTIONS: Sodium hypochlorite is household bleach – it may bleach and/or damage your clothing. Methanol and ethanol are flammable; avoid open flames.

Note: Procedures for the iodination of two different starting materials are provided. Your instructor will tell you whether you should carry out procedure A or procedure B.

A. Iodination of 4-hydroxyacetophenone to 4-hydroxy-3-iodoacetophenone

Reaction

1. In a round-bottom flask containing a magnetic stir bar, dissolve 1.0 g of 4-hydroxyacetophenone in 20 mL of methanol. To this solution, add 1.2 equivalents of potassium iodide, then cool to 0 °C with an ice/water bath.

2. Using a separatory funnel, add 1.0 equivalent of aqueous sodium hypochlorite solution (5.25% w/w) dropwise to the stirred reaction mixture over a period of 30 minutes.

3. Once the addition is complete, continue to stir the solution at 0 °C for 1 hour.

Workup and Isolation

4. Add 10 mL of sodium thiosulfate solution (10% w/w), then acidify with hydrochloric acid (10% w/w). The aryl iodide should precipitate at this point.

5. Remove the methanol from the suspension on a rotary evaporator. With gentle heating, this should require no more than 10 minutes. Note that water will still be present.

6. Cool the flask in ice for a few minutes, then collect the precipitate by vacuum filtration.

7. Recrystallize the crude aryl iodide from boiling water. A hot filtration is often required to remove 4-hydroxy-3,5-diiodoacetophenone, which is not soluble in water. Allow the hot solution to cool, then place the flask in an ice bath for a few minutes to ensure complete crystallization.

8. Collect the crystalline product by vacuum filtration and continue to draw air through it for several minutes to facilitate drying.

Characterization

9. Weigh the product and determine its melting point. (The literature value for the melting point is 155 – 156 °C [56].)

---

56. A. Sogawa, M. Tsukayama, H. Nozaki, and M. Nakayama, *Heterocycles*, **1996**, *43*, 101-111.

10. If you will be carrying out Experiment 13, save your product in a clean and suitably labeled vial.

B. Iodination of vanillin (4-hydroxy-3-methoxybenzaldehyde) to 5-iodovanillin (4-hydroxy-5-iodo-3-methoxybenzaldehyde)

Reaction

1. In a 100 mL round-bottom flask containing a magnetic stir bar, dissolve 1.0 g of vanillin in 20 mL of ethanol. To this solution, add 1.17 g of sodium iodide, then cool to 0 °C with an ice/water bath.

2. Using a separatory funnel, add 11 mL of aqueous sodium hypochlorite solution (5.25% w/w) dropwise to the stirred reaction mixture over a period of 10 minutes. The color of the solution will turn from pale yellow to red-brown.

3. Once the addition is complete, allow the mixture to warm to room temperature and continue to stir for 10 minutes.

Workup and Isolation

4. Add 10 mL of sodium thiosulfate solution (10% w/w), then acidify with hydrochloric acid (10% w/w). (Use pH paper to monitor the acidity; generally 6 mL or so of HCl solution are required.) The aryl iodide should precipitate at this point.

5. Remove the ethanol from the suspension on a rotary evaporator. With gentle heating, this should require no more than 10 minutes. Note that water will still be present.

6. Cool the flask in an ice bath for 10 minutes, then collect the precipitate by vacuum filtration. Wash well with ice-cold water, then with a small amount of cold ethanol. Continue to draw air through the crude product for several minutes to facilitate drying. Record the weight of the crude product.

7. If you will be using your product in Experiment 13, recrystallize the crude product from ethyl acetate. If you will not be using it in Experiment 13, you may instead recrystallize from aqueous 2-propanol: Place the crude product in a 100 mL Erlenmeyer flask and, with heating, add enough 2-propanol to dissolve it. While continuing to heat, gradually add hot water until the mixture becomes cloudy, then add enough 2-propanol to generate a clear (though colored) solution. Allow

186

the hot solution to cool, then place the flask in an ice bath for a few minutes to ensure complete crystallization.

8. Collect the crystalline product by vacuum filtration and continue to draw air through it for several minutes to facilitate drying.

Characterization

9. Weigh the product and determine its melting point. (The literature value for the melting point is 183 – 185 °C.)

10. If you will be carrying out Experiment 13, save your product in a clean and suitably labeled vial.

Post-Lab Questions and Exercises

1. Describe the physical properties (color and state) of your crude and purified product. Report the mass and percent yield. If your yield is low, suggest likely reasons.

2. How does your recorded melting point compare to the literature value? If there is a significant discrepancy, provide a plausible explanation for it.

3. What is the role of each reactant in this transformation? What is the role of the thiosulfate in the workup of the reaction?

4. Calculate the atom economy for the reaction.

5. Calculate your cost/gram of your product with and without considering the cost of waste disposal. Compare your cost to the cost per gram of your product from a commercial vendor.

Experiment Development Notes

This experiment presents an adaptation of a general iodination procedure reported in the primary literature by Edgar and Falling [57]. The iodination of 4-hydroxyacetophenone can be problematic and requires experimental care; diiodination, exacerbated by too-rapid addition of the sodium hypochlorite solution, appears to be the most significant problem. The vanillin iodination is appreciably more forgiving. These two reactions were selected in order to permit the use of the products in Experiment 13. If such a two-step procedure is not planned, the opportunity exists for student inquiry-driven investigation of other aromatic substrates for this iodination procedure.

57.   K. J. Edgar and S. N. Falling, *J. Org. Chem.* **1990**, *55*, 5287-5291.

EXPERIMENT 13

PALLADIUM-CATALYZED ALKYNE COUPLING/INTRAMOLECULAR ALKYNE ADDITION: NATURAL PRODUCT SYNTHESIS

Chemical Concepts

Alkyne chemistry (acetylides, addition reactions); substitution reactions; organometallic chemistry; air-sensitive compounds and techniques for handling them.

Green Lessons

Use of water as a solvent; catalysis; efficient synthetic routes.

Estimated Lab Time

3 hours (plus 0.5 - 1 hour preparation during preceding lab period)

Introduction

Terminal alkynes (R-C≡C-H) are quite acidic when compared with other hydrocarbons, and may be deprotonated by comparatively weak bases. The carbanion ("acetylide") resulting from deprotonation of a terminal alkyne can be used to effect the formation of a new carbon-carbon bond, either through nucleophilic addition to electrophilic reagents or through various other coupling reactions, such as the Glaser-Eglinton-Hay coupling reaction explored in Experiment 6. In this experiment, you will couple a terminal alkyne with an iodobenzene derivative in a reaction catalyzed by a palladium complex.

Palladium-catalyzed coupling of a terminal alkyne with an iodoarene

Alkynes can undergo a variety of addition reactions, including the addition of water or alcohols. The initial product of the alkyne coupling reaction in this experiment bears an OH group *ortho* to the alkyne functionality, and under the conditions of the coupling reaction, this OH group spontaneously adds across the carbon-carbon triple bond, forming a new five-membered ring. (This addition reaction may actually occur at an intermediate stage of the initial coupling reaction. In other words, addition of the OH group across the carbon-carbon triple bond may precede formation of the new carbon-carbon bond. This mechanistic complexity is significant, but for our purposes, the simpler reaction scheme put forth in this discussion is appropriate.) The isolated product of the coupling reaction is a benzofuran derivative. Interestingly, benzofurans have been isolated from a number of natural sources [58].

Intramolecular addition of an alcohol to an alkyne –formation of a benzofuran

This experiment presents two procedures, one starting with 4-hydroxy-3-iodoacetophenone, the other with 5-iodovanillin (4-hydroxy-5-iodo-3-methoxybenzaldehyde). The product arising from the former is a natural product, isolated from the giant tree daisy (*Podachaenium eminens*) [59]. The latter product has apparently not been found in nature, although the corresponding compound lacking the methoxy group is a natural product, isolated from a wood rot fungus (*Heterobasidion annosum*) that infects Sitka spruce trees (*Picea sitchensis*) [60]. (It appears very likely that the methoxy-containing derivative prepared in this experiment is also a natural product, but a natural source has apparently not yet been discovered.)

Benzofuran formation from 4-hydroxy-3-iodoacetophenone

58.   P. Cagniant and D. Cagniant, *Adv. Heterocycl. Chem.* **1975**, *18*, 337.
59.   G. E. Schneiders and R. Stevenson, *Synth. Commun.* **1980**, *10*, 699.
60.   D. M. Donnelly, N. Fukuda, I. Kuono, M. Martin, and J. O'Reilly, *Phytochemistry* **1988**, *27*, 2709.

Benzofuran formation from 5-iodovanillin

Whereas many reactions involving deprotonated alkynes are carried out in organic solvents with careful exclusion of water, in this experiment, you will use an aqueous solvent (either N-methylpyrrolidinone or 2-propanol) mixture. A water-soluble *ligand* (see Experiment 9) is required in order to dissolve the palladium catalyst in such a solvent mixture – here, we use a phosphine (phosphorus analog of an amine) bearing sulfonate groups, which impart high water solubility. A tertiary amine base – N-methylmorpholine – is added to take up hydrogen iodide as it is formed, and a catalytic amount of cuprous iodide is added to facilitate the reaction.

This experiment illustrates a number of green chemical principles. The reaction is effected by catalysis, affording the advantages discussed in Chapter 7. Although the reaction requires the presence of an organic solvent, benign solvents such as N-methylpyrrolidinone and 2-propanol are effective, and their use is minimized by their dilution with water. Finally, the reactions often proceed in high yield and are quite atom economical as well. This alkyne coupling/intramolecular addition reaction sequence has been reported to be successful for a variety of iodobenzene derivatives bearing adjacent OH or NHR (R = H, alkyl) groups, providing a relatively simple and general route to benzofuran and indole derivatives. The contrast of the mild and comparatively innocuous procedures and reagents used to effect the synthesis of these important heterocyclic compounds with the more traditional routes (involving lengthy reactions in refluxing pyridine or dimethylformamide) to these materials is striking.

Indole, a nitrogen-containing analog of benzofuran

Pre-Lab Preparation

1. Study the technique section in your lab manual regarding syringe use, thin-layer chromatography, extraction and washing, drying agents, and infrared spectroscopy.

2. Carry out pre-lab preparations as described in Chapter 11, section 11.6A, or as called for by your instructor.

3. Describe the role of each reagent used in the reaction.

4. 2-Methyl-3-butyn-2-ol, the alkyne used in this experiment, is a relatively inexpensive compound. Propose a simple one-step synthesis of this compound from any appropriate precursors containing three or fewer carbon atoms and any desired functional groups. (HINT: Think about a Grignard reagent.)

Experimental Procedure

> SAFETY PRECAUTIONS: 2-Methyl-3-butyn-2-ol is volatile and flammable and should be considered toxic as well. Handle with care, avoiding contact and inhalation. 3-Iodo-4-hydroxyacetophenone and 5-iodovanillin present no unusual safety hazards, but should be handled with care.

The coupling reaction is too slow to allow its completion in a single laboratory session. During the first session, you will measure out reagents and set up the reaction. In the second laboratory session, at least one day later, you will isolate, purify, and characterize the product.

This experiment requires the weighing out of a relatively large number of reagents. To ensure success, be careful in measuring the small amounts called for, and be sure to add *all* of the reagents. If you leave a reagent out, the reaction will not work.

Note: Follow either procedure A or B, as directed by your instructor.

A. Coupling of 4-hydroxy-3-iodoacetophenone with 2-methyl-3-butyn-2-ol

Reaction

1. Place 250 mg of 4-hydroxy-3-iodoacetophenone (Experiment 12) in a 20 mL scintillation vial. Add palladium(II) acetate (2 mole percent), trisodium tri(3-sulfonatophenyl)phosphine [3,3',3"-phosphinidyne-tris(benzenesulfonic acid) trisodium salt, or TPPTS, 4 mole percent], and cuprous iodide (5 mole percent).

2. Dissolve this mixture in 10 mL of a 4:1 mixture of water and N-methylpyrrolidinone (NMP), gently bubble N_2 through the solution for a few minutes, then seal the vial with a rubber septum. (It is important to minimize the amount of oxygen present during the reaction.)

3. Using a syringe, being sure to avoid air bubbles, inject 2.5 equivalents of N-methylmorpholine and 2 equivalents of 2-methyl-3-butyne-2-ol into the reaction mixture through the rubber septum.

4. Gently swirl the vial for a few minutes, doing your best to dissolve all of the solid material, and leave it in your drawer until the next lab period. The reaction is usually complete by the next day, but the extra reaction time will not cause any problems and sometimes affords improved yields.

Workup and purification

5. Add 10 mL of brine (saturated aqueous NaCl solution) and transfer the mixture to a 50 mL separatory funnel. Extract the aqueous layer twice with 10 mL portions of ethyl acetate. Wash the combined ethyl acetate extracts at least five times with 10 mL portions of water and once with brine. These repetitive water washes remove any remaining NMP (which is difficult to remove by evaporation), facilitating recrystallization.

6. Dry the ethyl acetate layer over anhydrous magnesium sulfate, then filter to remove the drying agent.

7. Remove the solvent on a rotary evaporator and recrystallize the crude product from petroleum ether. Collect the product by vacuum filtration, continuing to draw air through it for several minutes to allow it to dry.

Characterization

8. Weigh the product and determine its melting point. (The literature value for the melting point is 72 – 73 °C.) Record an infrared spectrum of the product in the form of a KBr pellet.

B. Coupling of 4-hydroxy-5-iodo-3-methoxybenzaldehyde (5-iodovanillin) with 2-methyl-3-butyn-2-ol

Reaction

1. Place 278 mg of 4-hydroxy-5-iodo-3-methoxybenzaldehyde (5-iodovanillin, Experiment 12) in a 20 mL scintillation vial containing a magnetic stir bar. Add 4.5 mg of palladium(II) acetate, 22.7 mg of trisodium tri(3-sulfonatophenyl)phosphine [3,3',3"-phosphinidyne-tris(benzenesulfonic acid) trisodium salt, or TPPTS], and 9.5 mg of cuprous iodide.

2. Add 10 mL of a 1:1 mixture of water and acetone, gently bubble N_2 through the solution for a few minutes, then seal the vial with a rubber septum. (It is important to minimize the amount of oxygen present during the reaction.)

3. Using a syringe, being sure to avoid air bubbles, inject 0.3 mL of N-methylmorpholine and 0.2 mL of 2-methyl-3-butyne-2-ol into the reaction mixture through the rubber septum.

4. Allow the reaction mixture to stir for 1 hour, then leave it in your drawer until the next lab period. The reaction is usually complete by the next day, but the extra reaction time will not cause any problems and sometimes affords improved yields. (If it is possible to stir continuously, yields are often improved, but this procedure works acceptably in the absence of such capability.)

Workup and purification

5. Transfer the mixture to a 100 mL separatory funnel, then add 20 mL of ethyl acetate, using some of this solvent to rinse the reaction vessel. **Note: Use special care throughout this step to avoid accidentally discarding the desired ethyl acetate layer.** Add 15 mL of water and 5-10 mL of saturated aqueous NaCl solution (brine), shake well, and separate the layers. Extract the aqueous layer with an additional 15 mL of ethyl acetate. Combine the two ethyl acetate extracts and wash twice with 15 mL portions of 10% aqueous HCl (removing N-methylmorpholine), then twice with

15 mL portions of water. Finally, wash twice with 15 mL portions of 10% aqueous NaOH solution (removing any unreacted starting material). A final wash with brine may help to remove excessive amounts of water.

6. Dry the ethyl acetate layer over anhydrous magnesium sulfate, then filter to remove the drying agent.

7. Remove the solvent on a rotary evaporator, then redissolve the crude product in a minimum amount of ethyl acetate. Filter the solution through a ca. 0.5-inch column of activated charcoal contained in a disposable pipette (using a small piece of glass wool as a plug to contain the charcoal), collecting the filtrate in a small Erlenmeyer flask.

8. Heat the filtrate to boiling and allow most of the ethyl acetate to evaporate. Add hot hexanes until the mixture becomes cloudy (10-15 mL), then add a few drops of ethyl acetate to clarify the solution, remove it from the heat source, and allow it to cool. If the product does not crystallize, cool the flask in ice and try scratching with a glass rod.

9. Collect the product by vacuum filtration, continuing to draw air through it for several minutes to allow it to dry.

Characterization

10. Weigh the product and determine its melting point. (The anticipated melting point is 108 – 110 °C.) Record an infrared spectrum of the product in the form of a KBr pellet.

Post-Lab Questions and Exercises

1. Describe the color and state of your purified product. Report the mass and percent of theoretical yield of the purified product. If your yield was low, provide a plausible reason(s) for the low yield.

2. Report the melting point *range* for your product.

3. Considering the need to exclude oxygen from the reaction, what oxidation state of palladium is likely to be the active species in this reaction? (You may need to consult a general chemistry text to determine the plausible oxidation states for palladium.)

4. Alcohols don't ordinarily react as readily with alkynes as seen in this experiment. Why might the alcohol (phenol) add so readily across the triple bond in this case?

5. Calculate the atom economy for the reaction.

6. Perform an economic analysis for the preparation of your product.

Experiment Development Notes

This experiment is based upon an original report in the primary literature by Amatore, *et al* [61]. The initial iteration of the experiment, producing the Spruce rot fungus natural product (a liquid), utilized aqueous N,N-dimethylformamide as the solvent. Use of the methyl ketone in place of the aldehyde afforded the solid natural product from the giant tree daisy and allowed the reaction to be carried out in aqueous NMP. The 5-iodovanillin reaction also affords a solid product, and it may be carried out in aqueous 2-propanol, not only greener but appreciably easier to remove than NMP. It may be possible to eliminate the phosphine entirely [62], and, if efficient stirring is available, sodium acetate may be used in place of N-methylmorpholine with sufficiently long reaction times (5-7 days). A variety of other terminal alkynes may be used for these coupling reactions; the original literature report provides several examples [61].

61.   C. Amatore, E. Blart, J. P. Genet, A. Jutand, S. Lemaire-Andoire, and M. Savignec, *J. Org. Chem.* **1995**, *60*, 6829-39.
62.   T. E. Goodwin, E. M. Hurst, and A. S. Ross, *J. Chem. Ed.* **1999**, *76*, 74.

EXPERIMENT 14

RESIN-BASED OXIDATION CHEMISTRY

Chemical Concepts

Oxidation chemistry; solid-supported reagents.

Green Lessons

Nontraditional reagents and conditions; recyclable reagents.

Estimated Lab Time

2 – 2.5 hours

Introduction

Oxidation and reduction represent nearly ubiquitous processes in the realm of synthetic organic chemistry. Given the central importance of such reactions, countless reagents and procedures have been developed to effect them. Typical oxidizing reagents are often corrosive, toxic, and environmentally damaging, and the development of environmentally benign procedures for the adjustment of oxidation state remains an important research goal. (The oxidation of cyclohexene in Experiment 5 provided an introduction to these issues.)

In this experiment, we will exploit the intrinsic reactivity and selectivity of a traditional oxidizing agent, CrO_3. Rather than the standard protocol of oxidation using CrO_3 in strong acid, however, we will use a "solid-supported" reagent, comprised of a polymer containing a reactive form of CrO_3 [63]. The polymer-supported CrO_3 readily oxidizes organic substrates, illustrated here with the oxidation of a secondary alcohol, 9-fluorenol, to the corresponding ketone, 9-fluorenone, but is retained in the solid support during and after the oxidation reaction. Not only does this render reaction work-up comparatively simple, but it also makes recovery of the heavy metal byproduct trivial. Even better, the polymer-supported byproduct may be regenerated and reused for further oxidation reactions, whereas the heavy metal products of simple oxidations are usually simply discarded.

---

63.   For a general discussion of the use of polymer-supported reagents, see: C. C. Leznoff, *Accts. Chem. Res.* **1978**, *11*, 327.

Oxidation of 9-fluorenol to 9-fluorenone

Pre-Lab Preparation

1. Study the technique section in your lab manual regarding heating reactions at reflux, thin-layer chromatography, rotary evaporators, recrystallization, and melting point determination.

2. Carry out pre-lab preparations as described in Chapter 11, section 11.6A, or as called for by your instructor.

3. The Aldrich catalog lists a "flash point" (Fp) for toluene of 4 °C. In this context, what is a flash point? (HINT: Reference to a book on chemical safety will probably lead you in the right direction.) Compare the flash point of toluene with those of two (or more) other common organic solvents (of your choosing).

4. It is common practice to mark the origin point in a TLC experiment before carrying out elution in order to make it easier to determine retention factors. Which of the following is an appropriate way to mark the origin before elution: pencil line, ink pen line, "magic marker" line, scratch? What would happen during elution if you used the method(s), if any, which you feel are inappropriate?

Experimental Procedure

SAFETY PRECAUTIONS: Toluene, acetone, hexanes, and ethanol are volatile and flammable solvents. Fluorenol and fluorenone present no unusual safety hazards, but should be handled with care.

The polymer-supported CrO_3 will be prepared for you. For completeness, the preparative details [64] for this reagent are presented here. With stirring, add 35 g of the chloride form of Amberlyst A-26 (macroreticular anion exchange resin with quaternary ammonium groups; Amberlyst A-29, Amberlite IRA 400, or Amberlite 904 may be used instead) to a solution of 15 g of CrO_3 in 100 mL of water. Stir for 30 minutes, then rinse with water, then acetone, then ether, and finally dry *in vacuo* at 50 °C for 5 hours. A typical "loading level" is ca. 3.8 mmol CrO_3/g of resin.

Reaction

1. Add 1 g of 9-fluorenol and 5 g of the dry polymer-supported CrO_3 reagent to 35 mL of toluene in a 100 mL round-bottom flask containing a magnetic stir bar.

2. Fit the flask with a water-cooled reflux condenser, then heat at reflux with magnetic stirring for approximately 1 hour.

3. Follow the progress of the reaction by thin layer chromatography on silica plates, eluting with 30% acetone in hexanes.

Workup and purification

4. Cool the reaction mixture to room temperature. Remove the polymer by filtration, rinsing with a small additional amount of toluene, then remove the solvent using a rotary evaporator. Place the used polymer in the recycling bottle provided.* You may find it necessary to use a gentle stream of air to remove the last traces of toluene from your crude product. Record the mass of the crude reaction product.

5. Recrystallize the crude product from ethanol or ethanol/water.

Characterization

6. Record the mass of the recrystallized product and its melting point.

*Note: The recovered resin retains a considerable amount of unreacted chromate ion and may be used several times without any additional treatment. If desired, it can be regenerated by treatment with fresh CrO_3 solution.

---

64.   G. Cainelli, G. Cardillo, M. Orena, and S. Sandri, *J. Am. Chem. Soc.* **1978**, *98*, 6737-6738.

Post-Lab Questions and Exercises

1. Report the mass and percent of theoretical yield of your crude product and of your recrystallized product, as well as the percent recovery in the recrystallization step. If your yield seems low, provide a plausible reason(s) for the low yield.

2. What was the melting point of your purified product? What is the reported melting point for 9-fluorenone? Provide a plausible explanation of any discrepancy between your melting point and that reported.

3. The active oxidizing agent in this experiment has been suggested to be [HCrO$_4$]$^-$. This relatively small anion looks like it should be freely soluble in water, yet the "solid-supported" version of this oxidizing agent is made by extracting it into the solid support from aqueous solution. What is it about the structure of the "ion exchange resin" used in this experiment that suggests it should hold strongly onto this anionic species?

4. Calculate the atom economy for the reaction.

5. Perform an economic analysis for the preparation of your product.

Experiment Development Notes

This chemistry forming the basis of this experiment was originally reported by Cainelli, *et al* [64]. An undergraduate experiment utilizing the supported oxidizing reagent was reported by Wade and Stell [65], and a variation appears in the classic microscale text by Mayo, Pike, and Trumper [66]. The procedure reported here represents a further modification, based on the primary literature report and local experience with the experiment.

65. L. G. Wade, Jr. and L. M. Stell, *J. Chem. Ed.* **1980**, *57*, 438.
66. D. M. Mayo, R. M. Pike, and P. K. Trumper, "Microscale Organic Laboratory," 3rd Ed., Wiley: New York, 1994, pp. 400-405.

EXPERIMENT 15

CARBONYL CHEMISTRY: THIAMINE-MEDIATED BENZOIN CONDENSATION OF FURFURAL

Chemical Concepts

Reactions of carbonyl compounds; oxidation chemistry; carbon skeleton rearrangements.

Green Lessons

Safer and easier to handle reagents and solvents.

Estimated Lab Time

2 hours (plus 0.5 – 1 hour preparation during preceding lab period)

Introduction

Compounds containing the carbonyl group display a surprisingly diverse range of reaction chemistry, and the discovery and study of such reactions form the heart of much modern synthetic organic chemistry. In the early days of organic chemistry, two chemists whose names are probably familiar to you, Friedrich Wöhler and Justus von Liebig, serendipitously discovered a cyanide-catalyzed reaction of benzaldehyde. In this reaction, two equivalents of benzaldehyde combine to form one equivalent of benzoin, containing a new carbon-carbon bond.

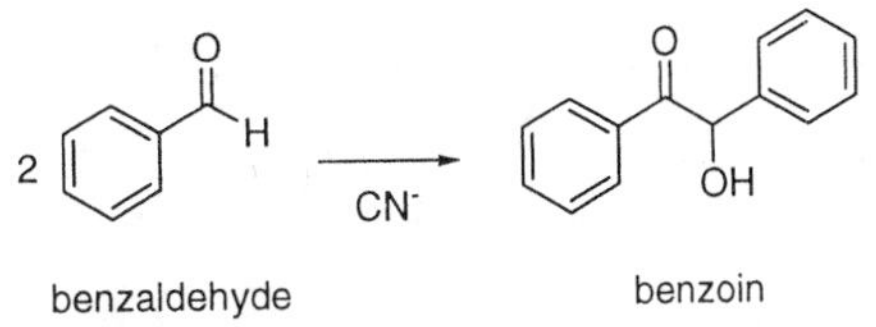

The benzoin condensation

This reaction, known as the benzoin condensation, represents an efficient way to construct carbon-carbon bonds and has been used to effect the synthesis of a number of interesting compounds. However, it was historically a frustrating reaction in that cyanide was essential to its success. The

201

cyanide salts typically used (e.g., sodium cyanide) are highly toxic, and inadvertent exposure to acidic conditions converts them to the volatile and toxic gas, hydrogen cyanide.

More than 100 years after the discovery of the benzoin condensation, it was found that, rather astoundingly, vitamin B_1 (thiamine) and related compounds could effectively catalyze the benzoin condensation. A partial mechanism for the thiamine-mediated reaction is provided below.

Thiamine hydrochloride

Mechanism of the benzoin condensation

This represents one of the earliest reports of the use of biologically derived reagents as tools to effect useful chemical transformations in the synthetic organic laboratory. Thiamine, a very safe compound, replaces the very dangerous cyanide salts previously required to effect the benzoin condensation. Much of the current heavy emphasis on the development of environmentally benign techniques for chemical synthesis is focused on the discovery of biologically assisted or inspired syntheses such as this thiamine mediated reaction. Topics under study include the use of "cofactors" (such as thiamine), the recruitment of the catalytic chemistry of purified enzymes, and the use of whole organisms (containing a wide range of enzymes and cofactors) to effect key chemical transformations.

The thiamine-mediated reaction is usually carried out using benzaldehyde. Another aldehyde, furan-2-carboxaldehyde (2-furaldehyde, furfural), however, has been reported to react similarly in the presence of thiamine [67,68]. In this experiment, we will explore the ability of thiamine to effect the formation of furoin from furfural.

The benzoin-type condensation of furfural

Pre-Lab Preparation

1. Study the technique section in your lab manual regarding heating reactions at reflux, recrystallization (including use of activated charcoal for decolorizing), and melting point determination.

2. Carry out pre-lab preparations as described in Chapter 11, section 11.6A, or as called for by your instructor.

Experimental Procedure

SAFETY PRECAUTIONS: Furfural (2-furaldehyde) is considered toxic. It is not extraordinarily volatile, but should be handled with care. Some byproducts may have objectionable odors, and working in the fume hood is advised. Sodium hydroxide solutions can cause severe skin and eye damage. Handle with extreme care, and avoid contact. Other reagents and intermediate products present no unusual safety hazards, but should be handled with care.

---

67.   R. Breslow, *J. Am. Chem. Soc.* **1958**, *80*, 3719.

The reaction is too slow to allow its completion in a single laboratory session. During the first session, you will measure out reagents and set up the reaction. In the second laboratory session, ideally one week later, you will isolate, purify, and characterize the product.

Reaction

1. In a 25 mL round-bottom flask equipped with a magnetic stirring bar, dissolve, with stirring, 0.30 g of thiamine hydrochloride in a mixture of 0.45 mL of water and 3.0 mL of 95% ethanol. At this point, the reaction mixture should be clear and colorless.

2. Dropwise, add 0.90 mL of a solution of 8.0 g of NaOH in 100 mL of water (this solution will be prepared for you). This step should resemble a titration – each drop of NaOH should produce a bright yellow color, which disappears with stirring. When all of the NaOH solution has been added, there should still be a pale yellow color remaining. If the reaction mixture is colorless, you must add more NaOH solution, or the reaction may not work.

3. Add approximately 0.73 mL of furfural, mix thoroughly, seal with a septum and store in your drawer until the next lab period.

Workup and purification

4. Cool the reaction mixture in an ice bath, adding several mL of water to the cooled mixture to drive the remainder of the product from solution, then isolate the crude solid product by filtration. Pull air through the solid on the filter to dry it, then record the mass of crude product obtained.

5. Recrystallize the crude product from 95% ethanol.

Characterization

6. Record the yield and melting point range of the purified material.

7. Obtain an infrared spectrum of your recrystallized furoin.

Post-Lab Questions and Exercises

1. Describe the physical properties (color and state) of your crude furoin and your recrystallized furoin. Report the mass and percent of theoretical yield of both the crude product and the purified

---

68. C. K. Lee, M. S. Kim, J. S. Gong, and I.-S. H. Lee, *J. Heterocyclic Chem.* **1992**, *29*, 149.

material, as well as the percent recovery in the recrystallization step. If your yield was low, provide a plausible reason(s) for the low yield.

2. Report the melting point range for your recrystallized furoin and the reported melting point of furoin.

3. Attach your infrared spectrum of furoin to your report. Identify the significant features of the spectrum, suggesting when possible vibrational assignments for the observed absorption bands.

4. Calculate the atom economy for the reaction.

5. Perform an economic analysis for the preparation of your product.

Experiment Development Notes

This procedure was adapted from the original reports by Breslow [67], Lee, *et al.* [68], and Hanson [69]. In contrast to analogous procedures for the preparation of benzoin, which we have found to be erratically reproducible, the furoin condensation generally works well. Furoin may be readily oxidized to the corresponding α-diketone, furil, using a variety of mild oxidizing agents, an example of which is hoped to be included in the next edition of this text. This could provide a convenient opportunity for independent student investigation.

69.   R. W. Hanson, *J. Chem. Educ.* **1993**, *70*, 257.

EXPERIMENT 16

SOLID-PHASE PHOTOCHEMISTRY

Chemical Concepts

Photochemistry; crystal engineering; recrystallization.

Green Lessons

Photochemical methods; solid-state (solvent-less) reactions; atom economy of addition reactions; selective synthesis; alternative energy sources.

Estimated Lab Time

1 hour (plus 0.25 hour preparation during preceding lab period)

Introduction

Cycloaddition reactions, in which two (or more) reactants come together with the concomitant formation of two (or more) new bonds, have proven particularly useful in organic synthesis, allowing the facile formation of multiple new bonds, often with good stereochemical control, in a single step. In this experiment, you will have the opportunity to explore a "simple" cycloaddition reaction, which is not really so simple in that 1) it requires light to occur (i.e., it is a photochemical reaction), and 2) it does not work at all if carried out the way most organic reactions are effected – by dissolving the reactants in a suitable solvent. Through these studies, you will gain experience in photochemistry and in the reaction chemistry of alkenes, focusing on the regiochemistry and stereochemistry of cycloaddition reactions. You will also have the opportunity to consider the structure and reaction chemistry of crystalline solids, concepts at the heart of the rapidly developing research field of "crystal engineering." In the course of your investigations, you will gain experience in the laboratory techniques of reflux, extraction, recrystallization, and melting point determination.

Cycloaddition reactions may be conveniently classified according to the number of electrons contained in bonds being broken in each of the reacting partners. For example, the well-known Diels-Alder reaction is referred to as a [4+2] cycloaddition reaction, since two pi bonds (containing four electrons) are broken in the *"diene"* reactant and one pi bond (containing two electrons) is broken in the *"dienophile"* (the alkene reacting with the diene).

diene   dienophile

The Diels-Alder ([4+2]) cycloaddition reaction

Many other cycloaddition reactions are also known. Perhaps the simplest conceptually is the [2+2] cycloaddition reaction, in which two alkenes combine to form a cyclobutane derivative.

The [2+2] cycloaddition reaction

Interestingly, however, while the Diels-Alder reaction is often effected at room temperature or, in more stubborn cases, with the application of heat, the [2+2] cycloaddition of simple alkenes generally fails. Analysis of the symmetry of the interacting molecular orbitals provides a convenient explanation of this marked difference in thermal reactivity. While such an analysis is beyond the scope of this discussion, a most intriguing outcome is that, whereas the thermal [2+2] cycloaddition reaction is unfavorable, the reaction of an alkene that has been electronically excited to a higher energy state is expected to be favorable. For typical alkenes, the energy required to cause a pi electron to be excited to a pi antibonding molecular orbital is provided by light in the visible to ultraviolet region of the electromagnetic spectrum, and thus these reactions are referred to as photochemical reactions. As expected, the photochemical [2+2] cycloaddition of alkenes is well known and can be a synthetically useful reaction.

The photochemical [2+2] cycloaddition reaction does have its limitations, however. A dramatic example is provided by *trans*-cinnamic acid (*trans*-C_6H_5-CH=CH-CO_2H), which could afford eleven possible products, six stereoisomers of the "head-to-head" dimer ("truxinic acids") and five of the "head-to-tail" dimer ("truxillic acids"). Both the truxinic acids and the truxillic acids are natural products, originally found in leaves of the coca plant (from which cocaine is derived) from Truxillo, Peru (from whence their names arise).

Possible products of the [2+2] cyclodimerization of trans-cinnamic acid

Curiously, the photochemical dimerization of *trans*-cinnamic acid is ineffective in solution. However, it does occur when crystalline *trans*-cinnamic acid is irradiated by visible light. Amazingly, this solid-state (solventless) reaction affords only a single stereoisomeric product.

Product of the solid-state photodimerization of trans-cinnamic acid

This experiment highlights a number of green chemical concepts. Since the cycloaddition reaction is carried out in the solid state, the issues of solvent choice, hazards, and disposal become moot. In addition, through the use of these rather unusual reaction conditions, a reaction that fails completely in solution is effected with reasonable efficiency, most importantly with the formation of a single stereoisomeric product. This synthetic efficiency is buttressed by the intrinsically high atom economy of a cycloaddition addition reaction, in which each atom of the starting material (in this experiment, cinnamic acid) is incorporated in the product of the reaction. Finally, the reaction may be effected with the energy of the sun rather than that of nonrenewable energy resources.

Pre-Lab Preparation

1. Study the technique section in your lab manual regarding melting point determination.

2. Carry out pre-lab preparations as described in Chapter 11, section 11.6A, or as called for by your instructor.

Experimental Procedure

SAFETY PRECAUTIONS: Cinnamic acid presents no unusual safety hazards, but should be handled with care. Tetrahydrofuran and toluene are volatile and flammable; avoid open flames and undue contact or inhalation.

This experiment is very quick to set up and to work up, but requires several weeks of irradiation.

Reaction

1. Place 1.5 g of *trans*-cinnamic acid in a 125 mL Erlenmeyer flask and add approximately 2 mL of tetrahydrofuran (THF).

2. Gently heat the mixture in order to dissolve the acid. Remove the flask from the heat source and gently rotate the flask while the solution cools in order to coat the walls of the flask with crystalline cinnamic acid. If the resulting coating is very uneven, reheat and try again. Once the coating is sufficiently dry, clamp the flask upside down for 30 minutes to allow the solvent vapor to escape.

3. Stopper the flask, label it with your name, and place it, upside down, in a beaker on a window ledge (with southern exposure if possible) or, if more appropriate, under a sunlamp. After 1 week, rotate the flask to expose the opposite side.

Workup and purification

4. After the second week of irradiation, transfer the solid to a 25 x 100 mm test tube or a small Erlenmeyer flask. Add 15 mL of toluene and warm to ca. 40 °C to dissolve any remaining cinnamic acid. Isolate the insoluble solid by filtration, rinsing with 10 mL of toluene, and continue to pull air through the product until it is dry.

Characterization

5. Air dry and weigh the product, and record the percent yield and melting point. (The literature value is 286 °C).

Post-Lab Questions and Exercises

1. Describe the physical properties (color and state) of your product. Report the mass and percent of theoretical yield of the product. If your yield was low, provide a plausible reason(s) for the low yield.

2. Report the melting point range for your product.

3. Draw the structures of the six diastereomers of the truxinic acids and the five diastereomers of the truxillic acids, clearly showing the stereochemistry. Present only one enantiomer for any chiral dimers, but bear in mind that in the photodimerization, chiral dimers are obtained as racemates.

4. Calculate the atom economy for the reaction.

5. Perform an economic analysis for the preparation of your product.

Experiment Development Notes

This experiment is based on reports in the primary literature and on an elegant experiment presented by Bell, *et al.* [70]. The latter presents this reaction as the first of a multi-step sequence of investigations designed to allow the determination of the structure of the photodimer by a combination of chemical and spectroscopic techniques. While the additional reaction chemistry required to deduce the structure does not illustrate any new green principles (and for that reason, is not included in this preliminary edition's presentation of the experiment), the sequence reported by Bell, *et al.* is very appealing.

70. C. E. Bell, A. K. Clark, D. F. Taber, and O. R. Rodig, "Organic Chemistry Laboratory: Standard and Microscale Experiments," Saunders College Publishing: Philadelphia; 2nd Ed., 1997, Ch. 32.

EXPERIMENT 17

APPLICATIONS OF ORGANIC CHEMISTRY:
PATTERNING SURFACES WITH MOLECULAR FILMS

Chemical Concepts

Organic materials chemistry; surface chemistry, self-assembled monolayers (SAMs), controlling surface properties with organic thin films; mechanical and chemical patterning of surfaces.

Green Lessons

Design of processes requiring less material ("dematerialization"); low-temperature, high-efficiency processes; benign solvents.

Estimated Lab Time

3 – 4 hours

Introduction

Although introductory organic chemistry lecture and laboratory courses often focus on the synthesis of small molecules, the modern practice of organic chemistry goes well beyond molecular synthesis. The design and preparation of new materials (solids, surfaces, polymers, etc.) displaying important properties represents an active area of research. When a thin film of an organic molecule is applied to the surface of a material, it is possible to change the surface properties of the material while maintaining its "bulk" properties. Organic thin films are used in a wide variety of applications. For example, coating an iron pan with a thin film of Teflon results in a non-stick cooking surface without significantly impacting the rate of heat transfer through the bulk iron of the pan. Other organic thin films are used as automobile waxes, anti-fogging treatments for windows and ski goggles, water- or stain-repellents for fabrics, and protective films for microelectronic components.

Use of organic thin films can be a useful strategy in designing greener products. For example, surface treatment can allow the preparation of products that require less use of cleaning materials given their stain-resistance or that have longer usable lifetimes given their protection from deterioration through corrosion. Use of thin films can also reduce the amount of material incorporated into a product, a strategy known as dematerialization, thereby reducing both the up-front chemical usage requirements for preparation of the product and the environmental impact of the product upon its disposal. The development of new coatings that provide desired surface properties, are easy to apply, and have long lifetimes is an area of ongoing research activity. In this experiment, you will explore the preparation of thin organic films on metal surfaces by a process known as *self-assembly*. As discussed below, in molecular self-assembly, the chemical functionality of molecules directs their organization into an extended structure or material. This method provides a convenient means of changing the surface properties of the metal – in this case, the hydrophobicity of the surface.

Pre-Lab Preparation

1. Study the following overview of self-assembled monolayer chemistry and the experimental section.
2. Carry out pre-lab preparations as described in Chapter 11, section 11.6A, or as called for by your instructor.

Self-Assembled Monolayers

Self-assembled monolayers (SAMs) are single molecular layers that spontaneously assemble on certain surfaces, driven by specific interactions between the surface and the monolayer-forming molecules. SAMs can be prepared using various types of molecules and surfaces. In this experiment, you will prepare and study SAMs of alkanethiols (RSH, the sulfur analogs of alcohols) on gold surfaces. A thiol-terminated molecule, typically containing 10-20 methylene units, is used to prepare alkanethiol SAMs. The sulfur "headgroup" binds strongly to the gold, creating a dense monolayer with the hydrocarbon

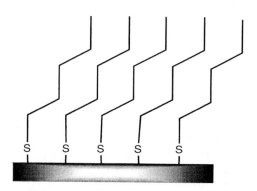

An alkanethiol monolayer on gold. The thin organic film is stabilized through binding of the sulfur headgroup to the surface and through van der Waals interactions between neighboring alkyl chains

"tails" pointing away from the surface. By using alkanethiol molecules with different tails, the surface chemical functionality can be varied widely, providing a simple way to change the functionality present on the surface.

Alkanethiol SAMs may be used in corrosion protection, in the preparation of electrically conducting molecular wires, and, as so-called photoresists, for the production of microelectronic devices. They can also serve as mimics of biological molecules or structures at surfaces and may be tailored to participate in designed interaction with desired molecules ("molecular recognition"). In each of these cases, the application of a *single-molecule-thick layer completely changes the surface properties of a material.* Thus, for example, a monolayer of an alkanethiol can completely inhibit the corrosion ("tarnishing") of a silver surface. In this experiment you will see how the hydrophobicity ("wettability") of a surface can be widely tuned using functionalized alkanethiol SAMs prepared from two molecules – one containing simply a hydrocarbon tail (hexadecanethiol), the other containing a carboxylic acid tail group (mercaptoundecanoic acid).

Preparation of Alkanethiol SAMs

Preparation of alkanethiol SAMs is generally straightforward. Most simply, a clean gold surface may be immersed in a dilute solution of the desired alkanethiol; formation of a monolayer occurs rapidly at room temperature. A variety of solvents are usable at the low thiol concentrations (typically 1-2 mM) that are used in preparation of SAMs; the most commonly used is ethanol.

Structures of Alkanethiol SAMs on Gold

SAMs have been thoroughly studied using a large number of analytical tools. SAMs prepared from 12-carbon or longer alkanethiols form well-ordered and dense monolayers on gold surfaces. The thiols are believed to lose the SH proton upon attaching to the gold surface, so that the surface is actually coated by a thiol*ate* monolayer. The alkyl chains are not oriented perpendicular to the surface, but rather are tilted at an angle of approximately 30° from the perpendicular orientation [71]. Well-formed, perfectly ordered SAMs can be envisioned as *two-dimensional crystals* of molecules on the surface.

---

71. The distance between binding sites for thiolates on the gold surface is around 5.0 Å, while the van der Waals diameter of an alkane chain is around 4.6 Å. By tilting, the alkyl chains come into closer contact with one-another.

Microcontact Printing with Alkanethiols on Gold

For many applications, it is necessary to pattern an organic thin film on a surface rather than form a complete film. In "microcontact printing," a rubber (elastomeric) stamp possessing the desired pattern is "inked" with the SAM-forming molecule. The inked stamp is brought into contact with a gold film, causing a SAM to form where the stamp contacts the surface. Once the stamp is removed, the film is placed in an aqueous chemical etching solution (iodine or cyanide-based etchants) that dissolves the

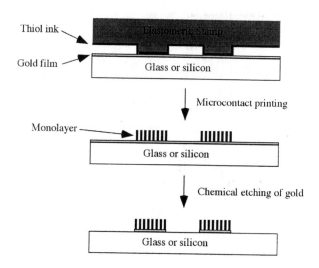

gold. The SAM acts as a chemically resistant layer, preventing the dissolution of the gold beneath the printed regions. The result after etching and rinsing is a patterned surface of gold. This technique can be used to pattern gold electrodes (contacts) and interconnecting wires on the surfaces of new electronic devices under development such as electronic paper and plastic transistors. (Interested students may wish to read the short section following this experiment for more detail about these applications.)

Chemical patterning of surfaces

Another patterning method, providing an alternative procedure to microcontact printing, takes advantage of the fact that the sulfur head groups of an alkanethiol SAM react with ozone to form water-soluble sulfonates. Following deposition of an alkanethiol SAM on a gold surface, the monolayer is covered by a "mask" that allows only the exposed portions of the SAM to react with ozone. Thus, the SAM is preserved beneath the mask, but removed (following ozone treatment and washing with water) everywhere else. The empty areas on the surface can be filled with a different thiol, leading to a monolayer patterned surface. Alternately, the unprotected gold can be dissolved through etching, leaving a gold patterned surface. In either case, this masking technology allows the creation of highly detailed patterns on the gold surface.

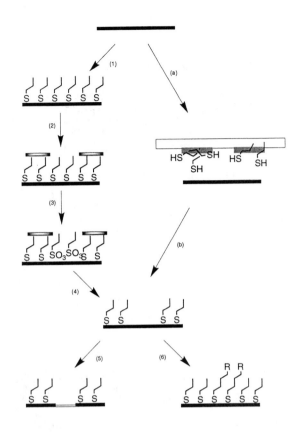

Reaction schemes for producing patterned SAMs on gold. Ozone patterning (left): (1) soaking the gold in a solution of alkanethiol to form a monolayer; (2) masking of part of the surface to protect that part of the film from ozone exposure; (3) exposure of the unprotected film to ozone, converting thiols to sulfonates; (4) removal of water-soluble sulfonates by washing, leaving a patterned film behind; (5) chemical etching of bare gold regions, or (6) backfilling of vacancies in the film with a different alkanethiol. Microcontact printing (right): (a) loading of thiol "ink" onto the stamp and (b) application to gold surface, forming a monolayer. The patterned monolayer can then be (5) etched or (6) backfilled.

Measuring contact angles

When a drop of liquid contacts a surface, it will either spread out across the surface (wet the surface) or bead up, depending upon the nature and strength of the interactions between the liquid and the surface. For example, rain falling on a freshly waxed automobile beads up, since the water molecules in a droplet interact more strongly with each other than with the waxed (hydrocarbon) surface. On the other hand, since glass surfaces contain many hydroxyl groups, water completely wets a clean drinking glass or window. The waxed surface is hydrophobic, while the clean glass surface is hydrophilic. We can assess the wettability of a surface by measuring the angle that a drop of water makes with the surface – the so-called contact angle. As the following diagram shows, a hydrophobic surface (like wax) has a high contact angle (Θ), while a hydrophilic surface has a low contact angle.

Examples of contact angle measurements of water on hydrophobic (left) and hydrophilic (right) surfaces. The contact angle θ is the angle between the substrate and a tangent to the droplet, taken at the point where the droplet meets the surface.

215

Contact angle values are easy to estimate and provide very useful information about the molecular-level composition and structural order on a surface. For example, the contact angle of a highly ordered alkanethiol SAM surface is about 118°, while a more disordered film, exposing some of the methylene groups of the alkyl chains, displays a lower contact angle (approaching 100°). If a SAM is comprised of two or more different alkanethiols, the film composition will affect the contact angle – a film that contains a higher proportion of hydrophilic tail groups will have a lower contact angle. You will measure the contact angles of your samples throughout this experiment to monitor your cleaning, assembly and patterning steps.

Experimental Procedure

> SAFETY PRECAUTIONS: Ethanol and acetone are volatile and flammable. When working with thiols (or solutions of thiols) gloves should be worn to reduce contact. Mercaptoundecanoic acid may be an irritant to eyes, respiratory system and skin. Although the stench associated with volatile thiols typically requires the use of a fume hood, the two thiols employed here are sufficiently non-volatile to be used on the bench top, if desired. All solvents and rinse water should be collected in separate waste containers and disposed of properly.

This laboratory exercise involves (i) the preparation and cleaning of the gold substrate, (ii) assembly and characterization of SAMs on gold, (iii) patterning monolayers through microcontact printing and ozone patterning, and (iv) visualization of monolayer patterns on surfaces. It is often convenient to work in small groups (pairs) during this experiment.

Preparation of gold on vinyl substrates

Gold films are often prepared using high-vacuum deposition equipment and can be quite expensive. The sign industry uses gold films on a vinyl backing, and this material represents a convenient and inexpensive source of gold films for the teaching laboratory. The gold film in such materials is typically protected with several plastic and adhesive layers, however, and these must be removed in order to expose the gold surface for monolayer formation. The procedure reported here works well for

SignGold (SignGold Corporation, 53 Smith Road, Middletown, NY 10941, USA), which requires removal of two polymer film layers and one layer of adhesive; use of other materials may require optimization of procedures for removal of coatings particular to those materials.

Each student (or student group) will need *three* gold film samples.

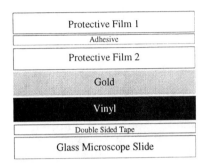

Cross-sectional representation of commercially available gold film on vinyl support attached to a glass microscope slide with double-sided tape. To expose the gold surface, the top two protective layers and the adhesive are removed, leaving a flat gold film suitable for monolayer formation.

1. Score and break a glass microscope slide to desired size. Three sections, each about 1 inch in length, work well.

2. Apply a single layer of double-sided tape to completely cover one side of the slide. Do not overlap tape or leave any of the glass showing on that side.

3. Press the tape-covered side of the slide onto the back (black surface) of the SignGold substrate. Apply pressure until the SignGold is firmly attached and as flat as possible. Press hard on the surface, and smooth any bubbles away by moving them toward the edge of the slide.

4. With the gold side of the assembly facing the bench top, use a sharp knife or razor blade to cut around the slide in order to remove any SignGold that extends past the edges. It is not necessary to cut all the way through the thick clear plastic film forming the top layer of the SignGold, since it will be removed in the next step. All areas of the surface of the slide should be covered by the substrate, and the substrate should not hang over the edges of the glass.

5. Turn the assembly over and remove the top, thick clear plastic film layer by carefully peeling it off of the assembly. This will reveal a sticky adhesive coating.

6. Soak the assembly in acetone for 3 - 5 minutes. The adhesive should start to dissolve, and the edges of the next polymer film layer should develop small ripples. Remove the slide from the acetone and wipe the surface, if necessary, to remove the adhesive.

7. Using tweezers, carefully remove the thin polymer film. Start at the ripples on an edge and pull the film back *parallel* to the surface of the substrate until it is completely removed. Perpendicular force may remove gold with the polymer film layer.

8. Rinse the gold surface with acetone and carefully wipe it clean.

Preparation of self-assembled monolayers

1. Prepare 10 mL of a 1 mM solution of hexadecanethiol in ethanol. Separately, prepare 10 mL of a 1 mM solution of mercaptoundecanoic acid in ethanol. It is not necessary to use great care in preparing these solutions, any concentration within a factor of two or so is fine. For the odoriferous hexadecanethiol, adding about 1 drop to 10 mL of ethanol is convenient.

2. Obtain three gold film samples, prepared as described above. Measure and record the water contact angle for one of the samples.

You will use contact angle measurements throughout this experiment to monitor your cleaning, assembly and patterning steps. To estimate the contact angle of a bare or monolayer-coated gold substrate, place a small (about 1 microliter) drop of Nanopure (or distilled) water on the surface. Estimate the angle Θ (see Figure at bottom of page 215) that the drop makes with the surface just at the point where the two meet. The easiest way to do this is to draw what you observe in your notebook and use a protractor to measure the angle.

You should observe an intermediate value for the contact angle, since the uncleaned gold surface is contaminated and not particularly hydrophobic or hydrophilic. No matter how your gold substrates are prepared, they will need to be cleaned just before use. Clean gold surfaces don't stay clean for long when they are exposed to air, quickly becoming coated with carbon-based contaminants. In order to form good SAMs on your gold films, they must be freshly cleaned. This is done by exposing them to ozone, generated by UV irradiation in the presence of oxygen in a UV/ozone Cleaner or a cleaning cabinet for goggles. Ozone oxidizes any organic contaminants adsorbed on the surface, converting them to water-soluble species that can be easily removed by rinsing with water.

3. Clean the gold films to prepare them for the self-assembly procedure. Place two of your substrates in the UV/ozone Cleaner. Start the cleaning process and leave the samples in the cleaner for 5 minutes. Remove the films and rinse each of them thoroughly with Nanopure (or distilled) water. Measure and record the contact angle of one of your cleaned films with pure water. If the film is clean, you should find a much lower contact angle than you recorded before the cleaning procedure. Once you have cleaned your samples, be especially careful not to contaminate them by touching them or spilling things on them. You should handle the gold surfaces only with clean metal tweezers.

4. Place one of the cleaned films in the hexadecanethiol solution and the other in the mercaptoundecanoic acid solution. Let them soak for at least 30 minutes, then remove your samples and rinse them thoroughly, first with ethanol and then with water. Blow away any excess water with dry nitrogen or clean compressed air.

5. Record the water contact angle for the two samples and note any other properties of the samples that you observe. For example, how does water run off the sample? Does the surface look different after formation of the monolayer?

Patterning of monolayers

1. Method A: Oxidative patterning with ozone.

Cover about half of the hexadecanethiol SAM coated slide with a small piece of a clean microscope slide, or cut a design out of the original protective film from the SignGold and lay it on the slide. (A detailed mask will produce clear patterns on the gold surface.) Place this slide in the UV/ozone Cleaner, along with the third, previously unused gold film sample (which will be used in Method B). Clean for 5-7 minutes (or longer if using a home-built cleaner), then remove the slides and rinse with Nanopure (or distilled) water. Soak the now-patterned hexadecanethiol SAM coated slide in the mercaptoundecanoic acid solution for 5 minutes, then rinse with Nanopure water and blow dry.

2. Method B. Contact printing.

Select a stamp (small rubber ink stamps available at crafts stores work fine) from those provided and "ink" it by coating the stamping surface with hexadecanethiol. Best results are obtained if excess hexadecanethiol is blotted off the surface by stamping momentarily on a paper towel or weighing paper. Bring the inked stamp into contact with the clean gold film (the film that has not yet been subjected to SAM formation) for a few seconds. If you see droplets of the thiol on the surface, you probably used too much. Immerse the film in the mercapto-undecanoic acid solution for about 5 minutes. Rinse with Nanopure water and blow dry.

3. Compare the properties of the two patterned surfaces using contact angle measurements, observing how water vapor condenses on the surface, how water sheets off the surface, how water pools on the surface, and any other things you can think of. The point is to explore ways to visualize your patterned surface and to compare the properties of these surfaces to those of the homogeneous films you prepared earlier.

Post-Lab Questions and Exercises

1. Draw a sketch and report the value for each of the contact angles that you observed (bare gold film before and after cleaning, $CH_3(CH_2)_{15}SH$ SAM, and $HS(CH_2)_{10}COOH$ SAM. Provide a brief explanation for the differences in contact angle that you observed.

2. Describe how you attempted to visualize the patterned SAMs. Which of the methods worked the best? Why? In retrospect, can you think of better ways to visualize the surface?

3. Describe the results of your patterning experiments. Did both of the methods work? Which do you think would be the better approach to patterning SAMs on a surface? Why would you make that choice?

4. Assuming that it is possible to make a drop of water move "uphill" on a surface (without external forces), how would you design the surface using SAMs to transport a drop of water uphill against gravity?

5. If a traditional photoresist is 1 - 2 micrometers in thickness, how much resist material is saved if a molecular film (such as those used in this lab) is used instead?

Experiment Development Notes

This is an original experiment developed at the University of Oregon, illustrating the use of monolayers to control surface properties. Although preparation of the gold substrates is time-consuming, they may be cleaned with ozone and water, assaying cleanliness with contact angle measurements, and then reused. Chemical etching following formation of patterned monolayers (discussed in the experiment but not included in the experimental procedures) often uses iodine or cyanide-based etchants. If greener etching conditions can be developed, actual patterning of the metal surface could be included in this experiment. It is possible to expand this laboratory exercise by presenting students with opportunities to solve problems such as making water move uphill (c.f. postlab question number 4). This experiment can also be modified for incorporation in the physical chemistry laboratory by including consideration of the Young equation.

Electronic Paper

The type of microcontact printing described above is the method used to pattern the gold electrodes and interconnects used in the Lucent/E Ink electronic paper. Each pixel is controlled by a novel type of transistor that is made up of organic chemical building blocks rather than the silicon-based building materials that make up traditional transistors. These plastic transistors are made using the process depicted in the Figure on the following page. Plastic substrates, organic insulating layers, alkanethiol patterning, and organic semiconductors are all examples of the use of organic materials in a device (a transistor) that has traditionally been comprised of inorganic materials such as silicon.

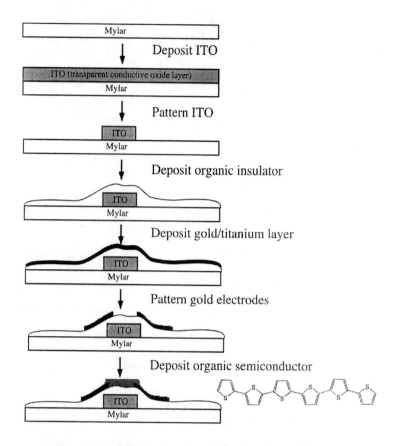

Construction of a transistor comprised largely of organic materials. A polymer, Mylar, is a flexible support for the device. An organic polymer is used as an insulating layer. A SAM is used to pattern electrodes, and an oligomer of thiophene acts as the active semiconductor in the device.

Notes for Instructors

This experiment works well with student teams (typically two students per team), and this can help to reduce waiting time for access to the ozone cleaners. If access to a thermal evaporator is available, evaporated gold films on glass substrates work very well. Gold-coated microscope slides are commercially available but prohibitively expensive. The experimental procedure describes use of commercially available and inexpensive gold films on vinyl backing. The preparation is somewhat time consuming, and it may be best to prepare the films in advance of the laboratory session. The following tips may prove helpful.

Preparation of gold films from SignGold

- When mounting the gold on vinyl on the glass substrate, make sure the double-sided tape is flat and does not overlap. Either overlap or space between the pieces of tape will allow acetone to reach the lower layers of the SignGold and wrinkle the vinyl from underneath.

- Remove air bubbles in the gold on vinyl assembly (attached to the glass slide) before removing either protective film. Pressing hard on the surface and smoothing the bubbles toward the edge of the slide seem to be the best ways to do this.

- Remember that there are *two* layers of protective plastic above the gold.

- It may be necessary to wipe the gold surface with acetone to remove all excess adhesive.

- Watch for the release of adhesive when soaking in acetone.

Monolayer formation

- Fresh alkanethiol solutions produce the best monolayers.

- Avoid contaminating the clean gold surfaces or thiol solutions. To avoid cross-contaminating the solutions, rinse slides with water before placing them into the next solution. Do not dip dirty tweezers in the solutions.

- Be sure to blot the rubber stamps for contact printing until the solution no longer spreads before stamping the gold surface.

Cleaning gold substrates with ozone

An effective and economical alternative to using a commercial UV/ozone cleaner is to build your own. A cleaner that works as well as a commercial unit can be constructed for less than $150. We purchased a Spa King UV/ozone Generator (a Bubble Gun II, Model # 6148, JED Engineering, Inc., 380 Vernon Way, Unit F, El Cajon, CA, USA 92020) from a spa retail store for $135 and used it with only minor modifications. A power cord was installed so that it could be plugged into a standard 110V wall outlet, and the screws were removed from the cover so that it could be taken on and off to place samples inside for cleaning. **Note:** *To prevent harmful UV exposure and the possibility of electrical shock, operate the unit only when the cover is in place.* This ozone generator can also be enclosed in a metal box equipped with a timer and a switch for the generator, as well as a drawer to hold the samples. We installed a safety mechanism that shuts the generator off if the drawer is opened during use.

When using either type of UV/ozone cleaner, the cleaning should be conducted in a fume hood due to the high concentrations of ozone produced by the generator (75-100 ppm). To use the home-built unit, place samples inside the ozone generator under the bulb, replace the cover and plug in (or switch on) the unit. In the Bubble Gun II, the tray below the bulb can accommodate 7 full size microscope slides or 21 1" samples at one time.

Contact angle measurements

Contact angles are best measured with very small drops. These drops can be produced with 1 μL pipette tips. Estimating contact angles as described in the experimental procedure, while sufficient for illustrating differences in wettability of methyl and carboxylic acid terminated monolayers, is not a very precise method. A goniometer is a device that can be used to more precisely measure contact angles. Although commercial contact angle goniometers are available, one that is reliable to within a few degrees can be constructed with a microscope, a mirror, a protractor and some wire. Place a 45° mirror on the base of the microscope, and observe the drop upside down. Align the crosshairs of the lens to the edge of the drop, and with a wire attached in alignment with the crosshairs, make a reading against a protractor.

EXPERIMENT 18

THE FRIEDEL-CRAFTS REACTION:

ACETYLATION OF FERROCENE

Chemical Concepts

Electrophilic aromatic substitution; recrystallization (hot filtration, use of decolorizing carbon); thin-layer chromatography; melting point determination.

Green Lessons

Safer reagents.

Estimated Lab Time

3 hours

Introduction

As has been noted in earlier experiments, a unifying theme of synthetic organic chemistry is the construction of carbon-carbon bonds. Such bond-forming reactions allow the elaboration of more complicated organic structures from simpler precursors. Experiment 12 introduced the concept of electrophilic aromatic substitution, in which an electrophile replaces a hydrogen atom in an aromatic compound. If the reactant in such a reaction is an electrophilic carbon species, a new carbon-carbon bond is formed. Such reactions, discovered in 1877 by Charles Friedel and James Crafts and now collectively known as Friedel-Crafts reactions, may be used to introduce both alkyl ("Friedel-Crafts alkylation") and acyl groups ("Friedel-Crafts acylation").

As the following mechanism indicates, Friedel-Crafts acylation is effected by the formation of an acylium ion as the active electrophilic species. The reactive acylium ion is generated from an acyl halide or anhydride by treatment with a Lewis acid; aluminum chloride is very commonly used for this purpose. Although $AlCl_3$ could potentially effect the catalysis of the Friedel-Crafts acylation reaction, the product, a ketone, is sufficiently basic to interact strongly with $AlCl_3$, so that in fact more than one equivalent of $AlCl_3$ is required. The $AlCl_3$ is removed in an aqueous workup step, which hydrolyzes it to HCl and aluminum hydroxide.

The Friedel-Crafts acylation reaction

Although many aromatic compounds are based on the prototypical benzene ring system, many other aromatic compounds are known. In general, aromaticity results from a cyclic, planar, fully-conjugated array of atoms with a total of 4n+2 (n = integer) pi electrons in the delocalized pi system. Thus, neutral compounds such as pyrrole and furan, as well as charged species such as tropylium, cyclopentadienyl, and cyclopropenium, all represent aromatic compounds.

Typical nonbenzenoid aromatic compounds

In this experiment, rather than using a simple benzene derivative as a reactant, you will explore the Friedel-Crafts acylation of a cyclopentadienyl ring contained in an organometallic compound (i.e., a compound containing one or more bonds between carbon and a transition metal). The substrate, ferrocene, contains two cyclopentadienyl rings that form a "sandwich" with the central iron atom [72]. Although the presence of the metal atom confers some unusual properties on ferrocene, the cyclopentadienyl rings undergo many reactions typical of aromatic compounds. In particular, ferrocene efficiently undergoes Friedel-Crafts acylation, and it is this reaction that you will investigate.

---

72. The discovery of ferrocene in 1951 heralded the arrival of the modern field of organotransition metal chemistry. From this relatively recent beginning, this field has grown enormously, and organotransition metal reagents and catalysts form integral parts of multi-billion dollar chemical industries, including the petrochemical and polymer industries.

226

Friedel-Crafts acylation of ferrocene, an organometallic complex containing cyclopentadienyl rings

Acylation of an aromatic ring deactivates it toward additional Friedel-Crafts acylation. However, since ferrocene contains a second cyclopentadienyl ring, it is possible for this second ring to be acylated, forming diacetylferrocene. The two cyclopentadienyl rings in ferrocene do "communicate" with one another electronically, so that introduction of the second acyl group requires somewhat more forcing conditions. Nonetheless, the effect is relatively subtle, and mixtures of mono- and diacylated ferrocenes are often obtained.

diacetylferrocene

Friedel-Crafts acylation represents a powerful and effective way to introduce new carbon-carbon bonds in aromatic compounds, and it has been extensively exploited as a synthetic tool since its discovery. However, the reaction is not without its limitations. A strong Lewis acid, often aluminum chloride, which is corrosive and gives off HCl upon contact with moist air, is required in greater than stoichiometric amounts, leading to the generation of considerable quantities of acidic and aluminum-containing waste. Common solvents for Friedel-Crafts acylation reactions include halogenated methanes (e.g., dichloromethane) or carbon disulfide, representing environmental and/or human health risks. In this experiment, you will use a more benign catalyst, phosphoric acid, to effect the Friedel-Crafts acylation reaction. No organic solvents are used, although one of the reactants, acetic anhydride, is used in excess and thus may play the role of a solvent in this reaction. Unfortunately, these acylation conditions are not general – it is the relatively high reactivity of ferrocene, compared to simpler aromatic substrates, which allows the replacement of $AlCl_3$ with phosphoric acid. Discovery of new Friedel-Crafts-like reaction chemistry, applicable to simple benzene derivatives, remains an area of ongoing investigation.

Pre-Lab Preparation

1. Study the technique sections in your lab manual regarding thin-layer chromatography, recrystallization (including hot filtration and the use of decolorizing charcoal), and melting point determination.

2. Carry out pre-lab preparations as described in Chapter 11, section 11.6A, or as called for by your instructor.

Experimental Procedure

> SAFETY PRECAUTIONS: Phosphoric acid and acetic anhydride are corrosive, and acetic anhydride is also a lachrymator; avoid contact or undue exposure to vapors.

Reaction

1. Place 1.5 g of ferrocene in a 20 mL round-bottom flask containing a magnetic stir bar. If a steam bath is not available, prepare a hot water bath, heating the water to nearly the boiling point while preparing the following reaction mixture.

2. In a fume hood, add 5.0 mL of acetic anhydride and 1.0 mL of 85% phosphoric acid to the flask. The reaction mixture should heat up and darken in color. Swirl the flask, heating occasionally in a hot water bath, if necessary, until all the ferrocene dissolves.

3. Attach a reflux condenser equipped with a calcium chloride drying tube, then heat the reaction mixture, with stirring, on a steam bath or in the hot water bath prepared in step 1. Heat for 10 minutes, during which time a purple color may develop.

Workup and purification

4. Pour the reaction mixture onto 25 g (ca. 60 mL) of ice in a 200 mL beaker, rinsing the flask with two 5 mL portions of ice water. (A black residue may remain in the flask.) Stir the orange-brown mixture with a glass rod for a few minutes. Any insoluble black material present will be removed in the following steps.

5. Add 37.5 mL of 3M aqueous NaOH solution, then *carefully* add solid sodium bicarbonate in small portions until the remaining acid has been neutralized (about 7 - 8 grams). (Use great care to avoid excessive foaming during this bicarbonate addition. This step can be done with magnetic stirring, but make sure to use a stirring plate that is not hot.) Stir well and crush any lumps, affording a dark-brown suspension.

6. Allow the mixture to stand for 20 minutes, then collect the crude product by vacuum filtration and continue to pull air through the product for a few minutes to dry it. Finish the drying process by pressing the solid product between two sheets of filter paper or paper towels. Save a bit of this crude product for TLC analysis.

7. Transfer the solid and a stir bar to a small Erlenmeyer flask and add 20 mL of hexanes. Boil for 5 minutes with stirring, then decant the dark-orange solution into another Erlenmeyer flask, leaving behind a black gummy substance.

8. To the hot solution, add a spatula-full of decolorizing carbon. (Use of too much carbon will reduce your yield.) Heat with swirling, then perform a hot filtration to remove the decolorizing carbon.

9. Set the flask aside to cool slowly. Red-brown needles of acetylferrocene should begin to form. Once the flask has reached room temperature, cool it in ice. Collect the crystalline product by vacuum filtration, washing with a small quantity of cold hexanes, and dry it by continuing to pull air through it for a few minutes.

Characterization

10. Record the yield and melting point range of your recrystallized acetylferrocene. (The melting point has been reported as either 82 - 83 or 84 – 85 °C.)

11. Analyze your crude and recrystallized products by TLC. Separately dissolve very small amounts of pure ferrocene, the crude product, and the recrystallized acetylferrocene in a few drops of toluene. Spot the solutions on silica gel plates and develop with 30:1 toluene/absolute ethanol. Visualization is simple – each of the compounds is brightly colored.

Post-Lab Questions and Exercises

1. Describe the physical properties (color and state) of your crude product.

2. Report the color and melting point range of your recrystallized product. Report the mass and percent of theoretical yield of the recrystallized product.

3. Report the results of your TLC analyses, including R_f values and discussing any differences between the crude and recrystallized products.

4. Calculate the atom economy for the reaction.

5. Perform an economic analysis for the preparation of your product.

Experiment Development Notes

This experiment represents an adaptation of a procedure reported by Fieser and Williamson [73]. The contrast of the reaction conditions called for in this experiment with those of more conventional Friedel-Crafts acylation reactions is striking, and if conditions allow, it may be instructive to allow students to explore this contrast experimentally. In addition to the differences in solvents and reagents, students carrying out such explorations should note that other experimental procedures for the acylation of ferrocene, using more vigorous acylation conditions, tend to afford mixtures of mono- and diacylated products.

73. L. F. Fieser and K. L. Williamson, "Organic Experiments," 8th Edition; Houghton Mifflin Co: New York, 1998, p. 332-333.

EXPERIMENT 19

COMBINATORIAL CHEMISTRY:

ANTIBIOTIC DRUG DISCOVERY

Chemical Concepts:

Combinatorial chemistry; carbonyl chemistry; antibiotics; sterile techniques; biological assays.

Green Lessons:

Synthetic efficiency; design of safer products; waste minimization.

Estimated Lab Time: 2 hours

Introduction

Drug development is a very expensive and time-consuming business. The initial process of finding a suitable lead compound may involve the screening of thousands of different compounds, looking for the desired activity, and the further development of drug candidates once an initial lead compound has been identified may require even more. Each of these compounds must be synthesized, purified, and tested. Recently, a fundamentally new approach has radically changed the nature of drug discovery and development. Using the procedures of "combinatorial chemistry, "tens to hundreds or even thousands of compounds are synthesized simultaneously in a single reaction. The resulting "library" of compounds is then tested for the desired activity, and promising candidates are then chosen for independent synthesis and more comprehensive testing. A number of variations on this theme have arisen, differing as to the specifics of library synthesis and activity testing. In one of the simpler approaches, the collective library is tested. If it proves active, the library is "deconvoluted" to identify the specific compound responsible for the activity. This can be done by various means, including re-synthesizing subsets of the initial pool, spatially arranging components of the library on a solid support during synthesis, or cross-indexing between different mixtures that contain the active compound. The latter approach will be clarified for you as you carry out this laboratory investigation, which is based on an exquisite report by Wolkenberg and Su [74].

---

74.   S. E. Wolkenberg and A. I. Su, *J. Chem Ed.* **2001**, *78*, 784-785.

In this experiment, you will prepare small libraries of hydrazones from various mixtures of aldehydes and hydrazines, according to the following general reaction.

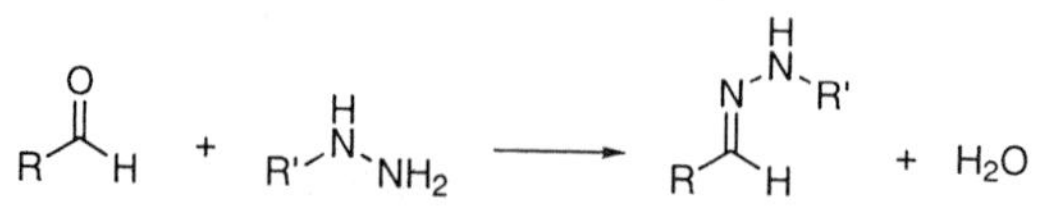

Condensation of an aldehyde with a hydrazine forms a hydrazone

By combining various mixtures of four aldehydes, labeled *A1 – A4*, and four hydrazines, labeled *B1 – B4*, you will form a total of sixteen different hydrazones, labeled *A1–B1*, *A1–B2*, etc., in eight mixtures, each containing four hydrazones. The aldehydes and hydrazines to be used are presented below.

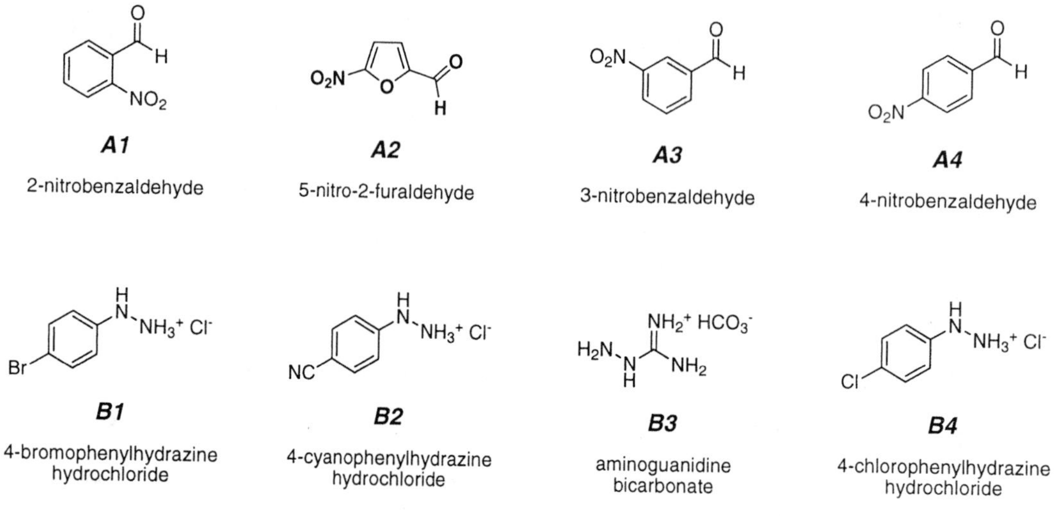

A1

2-nitrobenzaldehyde

A2

5-nitro-2-furaldehyde

A3

3-nitrobenzaldehyde

A4

4-nitrobenzaldehyde

B1

4-bromophenylhydrazine
hydrochloride

B2

4-cyanophenylhydrazine
hydrochloride

B3

aminoguanidine
bicarbonate

B4

4-chlorophenylhydrazine
hydrochloride

Reactants for combinatorial synthesis

You will screen the resulting mixtures of hydrazones for antibiotic activity against a common bacterium, *Eschericia coli* (*E. coli*) using a technique known as the "cup agar diffusion method." This simply involves the growth of a colony of *E. coli* on an agar plate containing small holes, each of

232

which contains a different mixture of hydrazones. If a mixture contains at least one active hydrazone, it will prevent bacteria from growing near the hole containing that mixture. By analyzing which active mixtures contained which specific hydrazones, you will be able to pinpoint the exact compound responsible for any observed antibiotic (antibacterial) activity.

The techniques of combinatorial chemistry are applicable to the discovery of compounds displaying other properties than drug activity as well. Indeed, the property of focus can be almost anything – electrical conductivity, appealing color or fragrance, prevention of pollen binding to a histamine receptor, destruction of cancerous cells or bacteria. One great advantage of combinatorial chemistry over traditional synthetic methods is that it decreases the number of preparative-scale independent syntheses required to identify promising compounds. In addition to saving time and effort, this can significantly reduce the generation of hazardous waste and reliance on hazardous reagents, reactants, and solvents.

Pre-Lab Preparation

1. Study the experimental procedure carefully.
2. Carry out pre-lab preparations as described in Chapter 11, section 11.6A, or as called for by your instructor.

Experimental Procedure

> SAFETY PRECAUTIONS: Laboratory strains of *E. coli* are non-toxic, but it is still a good idea to be careful. Wear gloves and goggles, and avoid touching your face or any scrapes/cuts with your gloves. Wipe down the bench top when you leave with the disinfectant provided, and wash your hands and arms. Dispose of all waste in the appropriate designated containers. Some of the chemicals used in this lab are considered toxic when used neat; however you will use them at very low concentrations.

Note: It is important to keep the Petri dishes covered at all times. Reduce the risk of contamination by being careful to cover the plates when not in use.

Setup and Preparation

1. Obtain three sterile agar plates. If these have not been prepared for you, prepare them by pouring heated agar solution [75] into the plates, filling them approximately half full. Quickly cover and allow to cool.

2. Once the agar plates are cool and firm, turn two plates over and use a marker to draw four divisions on the bottom of each plate. Number each division *M1*, *M2*, *M3*, *etc.* for the eight mixtures that will be used and write your name or initials on the bottom of each plate to identify it. The third plate, unlabeled, will be used in step 5 to practice preparing "cups" to hold the mixtures.

3. Turn the two labeled plates back over so that the tops are up. (*Do **not** add bacteria to the third, unlabeled plate.*) Remove the lids and pour approximately 5 mL of the *E. coli* culture (prepared in advance by the instructor according to the procedure reported at the end of this experiment) onto the first plate. Cover and swirl gently so that the bacteria evenly coat the agar surface. Tip the plate and transfer the excess liquid by pipette to the second plate. Cover and swirl gently so that the bacteria evenly coat the agar surface. Tip the plate and remove the excess liquid by pipette. Dispose of this liquid in the appropriate waste bottle and the pipette in a biohazard disposal container.

4. Using the practice (unlabeled) agar plate and a fresh pipette, practice making holes in the agar by pushing the large end of the pipette straight down into the agar and then pulling it straight up. Carefully remove the resulting cylinder of agar with the small end of the pipette or with a spatula, and place them in the biohazard disposal container.

5. Once you have perfected the technique, create one hole in each sector on your two agar plates containing *E. coli*. Replace the covers as soon as you are done.

---

75. Prepare the agar solution from 40 g of dehydrated LB agar and 950 mL of distilled water, sterilized by autoclaving for 20 minutes at 15 psi on liquid cycle.

Reaction

6. Obtain eight test tubes. Label them *M1, M2, M3, etc.* for the eight mixtures that you will prepare.

7. To each test tube, add the reagent solutions (prepared by the instructor in advance according to the procedure reported at the end of this experiment) *in the correct amounts and order* as called for below. Be sure to use a different pipette for each solution to avoid cross-contamination. If a reagent solution appears cloudy or if a precipitate has formed, heat it gently in a hot water bath for a few minutes and shake well. Once the precipitate has dissolved, the solution may be used.

To prepare mixtures *M1-M4*, place 5 drops of *B1* in each of four test tubes. Add 5 drops of *B2* to each tube, then 5 drops of *B3*, then 5 drops of *B4*. Finally, add 20 drops of *A1* to the first tube to prepare *M1*, 20 drops of *A2* to the second tube to prepare *M2*, 20 drops of *A3* to the third tube to prepare *M3*, and 20 drops of *A4* to the fourth tube to prepare *M4*. Make note of anything interesting that happens during these additions. Cap each tube, shake it for 10 seconds, and again make note of any significant changes.

To prepare mixtures *M5-M8*, place 5 drops of *A1* in each of four test tubes. Add 5 drops of *A2* to each tube, then 5 drops of *A3*, then 5 drops of *A4*. Finally, add 20 drops of *B1* to the first tube to prepare *M5*, 20 drops of *B2* to the second tube to prepare *M6*, 20 drops of *B3* to the third tube to prepare *M7*, and 20 drops of *B4* to the fourth tube to prepare *M8*. Make note of anything interesting that happens during these additions. Cap each tube, shake it for 10 seconds, and again make note of any significant changes.

Screening

8. Using a clean pipette for each mixture, carefully add one or two drops of each mixture, *M1-M8*, to the appropriately labeled hole on the agar plates. Make note if you spill any reagents on top of the agar or put a mixture in the wrong hole, as this will affect your interpretation of the results of the experiment.

9. Cover and stack your plates, placing them in the designated location. Be sure not to spill any liquid out of the holes while carrying your plates. The plates will be incubated overnight at 37°C.

10. Return to the lab the next day to check your results, recording your observations in your laboratory notebook. For each division on the plate, decide whether or not there was any growth of the bacteria. Make the answer a simple 'yes' or 'no'. You may have some mixtures with low activity, but we are looking for maximum activity.

Post-Lab Questions and Exercises

1. Which of the mixtures *M1-M8* appeared to inhibit significantly the growth of *E. coli*?

2. Complete the following table illustrating the composition of each of the mixtures *M1* to *M8*. Several entries have been made to illustrate how to do this.

mixture		M1	M2	M3	M4
	components	A1	A2	A3	A4
M5	B1	A1-B1			
M6	B2				
M7	B3			A3-B3	
M8	B4				

3. Shade the column corresponding to the mixture *M1-M4* that displayed antibacterial activity. Shade the row corresponding to the mixture *M5-M8* that displayed antibacterial activity. The intersection point is the active compound. What is the structure of this compound?

4. How may reactions and antibiotic screenings would have been required to carry out this analysis in a non-combinatorial fashion? If we were interested in examining 50 different aldehydes and 50 different hydrazines, how many reactions and antibiotic screenings would be required to carry out this analysis in a non-combinatorial fashion? How many using the combinatorial technique introduced in this experiment?

5. Calculate the atom economy for the formation of hydrazone *A1-B1* from aldehyde *A1* and hydrazine *B1*.

Experiment Development Notes

This experiment represents an expansion of that reported by Wolkenberg and Su [74]. We urge consultation of this original report for complementary discussions. We have added a fourth aldehyde (4-nitrobenzaldehyde) and a fourth hydrazine (4-chlorophenylhydrazine) to increase the size of the antibiotic libraries.

Notes for Instructors

It is convenient to have students work with partners for this experiment. Wolkenberg and Su [74] provide excellent discussions, and their publication should be consulted for additional information and details.

Preparation of the E. coli culture

Wolkenberg and Su [74] report a detailed procedure for *E. coli* culture growth. We have found the following simplified procedure, using an *E. coli* culture kit purchased from Carolina Biological Supply Co., to be sufficient. Reconstitute the *E. coli* according to the instructions provided, then add this culture to the desired quantity of sterile Luria-Bertani (LB) broth. [The LB broth may be purchased (Carolina Biological Supply Co.) or prepared as follows. Add 2.5 g of dehydrated LB broth (Fisher Scientific) to 95 mL of deionized water, then autoclave for 20 minutes at 15 psi on liquid cycle.] Cap, the store at 37 °C for 48 hours. When working with 200 mL of LB broth, this provides an appropriate concentration of *E. coli*; if larger quantities of broth are used, a longer incubation period may be appropriate. We have found it convenient to transfer ca. 5 mL portions of this culture to individual sterile containers for each student team, rather than requiring each student to work with the full culture.

Preparation of the reagent solutions

Solution *A1*: Place 54 mg of 2-nitrobenzaldehyde and 12 mL of deionized water in a sealable container (e.g., a centrifuge tube). Seal the container, then heat in a boiling water bath (or on a steam bath). With shaking, the compound should dissolve. Store the solution at room temperature until needed. If solid precipitates, redissolve it by heating.

The other reagent solutions should be prepared analogously, using the following quantities.

A2: 51 mg of 5-nitro-2-furaldehyde

A3: 54 mg of 3-nitrobenzaldehyde

A4: 54 mg of 4-nitrobenzaldehyde

B1: 80 mg of 4-bromophenylhydrazine hydrochloride

B2: 61 mg of 4-cyanophenylhydrazine hydrochloride

B3: 49 mg of aminoguanidine bicarbonate

B4: 64 mg of 4-chlorophenylhydrazine hydrochloride

Appendix A: The Twelve Principles of Green Chemistry

This text approaches the development of green chemical processes by considering a generic chemical equation and the ways in which one may reduce the hazards and environmental impacts arising from each component of this equation (Chapter 5). Anastas and Warner have formulated twelve fundamental principles of green chemistry, and these principles, as put forth in P. T. Anastas and J. C. Warner, "Green Chemistry: Theory and Practice;" Oxford University Press: Oxford, UK (1998), are reproduced below. Although we have not organized the discussions of this text around these principles, the concepts contained within each should sound familiar to you.

1. It is better to prevent waste than to treat or clean up waste after it is formed.

2. Synthetic methods should be designed to maximize the incorporation of all materials used in the process into the final product.

3. Wherever practicable, synthetic methodologies should be designed to use and generate substances that possess little or no toxicity to human health and the environment.

4. Chemical products should be designed to preserve efficacy of function while reducing toxicity.

5. The use of auxiliary substances (e.g. solvents, separation agents, etc.) should be made unnecessary wherever possible and innocuous when used.

6. Energy requirements should be recognized for their environmental and economic impacts and should be minimized. Synthetic methods should be conducted at ambient temperature and pressure.

7. A raw material or feedstock should be renewable rather than depleting wherever technically and economically practicable.

8. Unnecessary derivatization (blocking group, protection/deprotection, temporary modification of physical/chemical processes) should be avoided whenever possible.

9. Catalytic reagents (as selective as possible) are superior to stoichiometric reagents.

10. Chemical products should be designed so that at the end of their function they do not persist in the environment and break down into innocuous degradation products.

11. Analytical methodologies need to be further developed to allow for real-time, in-process monitoring and control prior to the formation of hazardous substances.

12. Substances and the form of a substance used in a chemical process should be chosen so as to minimize the potential for chemical accidents, including releases, explosions, and fires.

Index

241